Oxford Applied Mathematics and Computing Science Series

Oxford Applied Mathematics and Computing Science Series

I. Anderson: *A First Course in Combinatorial Mathematics (Second Edition)*

D. W. Jordan and P. Smith: *Nonlinear Ordinary Differential Equations (Second Edition)*

B. Carré: *Graphs and Networks*

G. D. Smith: *Numerical Solution of Partial Differential Equations (Third Edition)*

A. B. Tayler: *Mathematical Models in Applied Mechanics*

R. Hill: *A First Course in Coding Theory*

P. Baxandall and H. Liebeck: *Vector Calculus*

P. Thomas, H. Robinson, and J. Emms: *Abstract Data Types: Their Specification, Representation, and Use*

D. J. Acheson: *Elementary Fluid Dynamics*

L. M. Hocking: *Optimal Control: An Introduction to the Theory with Applications*

S. Barnett: *Matrices: Methods and Applications*

O. Pretzel: *Error-Correcting Codes and Finite Fields*

D. C. Ince: *An Introduction to Discrete Mathematics, Formal System Specification, and Z (Second Edition)*

A. Davies and P. Samuels: *An Introduction to Computational Geometry for Curves and Surfaces*

P. Grindrod: *The Theory and Applications of Reaction-Diffusion Equations: Patterns and Waves (Second Edition)*

O. Pretzel: *Error-Correcting Codes and Finite Fields (Student Edition)*

PETER GRINDROD
QuantiSci

The Theory and Applications of Reaction-Diffusion Equations

Patterns and Waves

Second Edition

CLARENDON PRESS · OXFORD
1996

Oxford University Press, Walton Street, Oxford OX2 6DP

Oxford New York
Athens Auckland Bangkok Bombay
Calcutta Cape Town Dar es Salaam Delhi
Florence Hong Kong Istanbul Karachi
Kuala Lumpur Madras Madrid Melbourne
Mexico City Nairobi Paris Singapore
Taipei Tokyo Toronto
and associated companies in
Berlin Ibadan

Oxford is a trade mark of Oxford University Press

Published in the United States by
Oxford University Press Inc., New York

A catalogue record for this book is available from the British Library

Library of Congress Cataloging in Publication Data
(Data available)

ISBN 0 19 850004 1

Typeset by the author

Printed in Great Britain by
Bookcraft Ltd
Midsomer Norton, Avon

To Dora

Preface to second edition

Since the first edition of this book came out I have become more and more aware of applications of reaction-diffusion models to problems within the environmental sciences in general, and to the geochemistry of water–rock interactions in particular. Whilst many of the analytical and conceptual tools within this book can carry straight over to these problems, I felt it was important to add a new chapter. This deals with geochemical systems, which tend to have a very large number of variables and almost instantaneous reactions, and propagating waves which are flux driven rather than excitable.

I am indebted to Dora, Tommy, Chris, Jumbly and Sophie for the time I stole from them in order to complete this task. I also want to thank Mark Penney for typesetting Chapter 6.

Henley-on-Thames P.G.
December 1995

Preface to first edition

To say that this is still exactly the book I set out to write is untrue. To say that it has come close will have to suffice. My aim has been to produce a text that contains as many constructive techniques and ideas as I could reasonably fit in. Necessarily, I have of course been selfish and selected topics and examples which suit my own preference rather than attempt to write a general survey of techniques and applications.

I hope that I have been able to communicate some of the sense of fun that I gain from my own modelling and applied maths. Although analysis has a serious, rigorous side, it is often driven by the particular application and we must be ever alert in order to have the right tool for the job in hand.

I have been helped along throughout my career by a number of people who have all had the time and patience to let me breathe as well as work. Among them, Brian Sleeman, Elias Tuma, Jim Murray, Jagan Gomatam and Yuzo Hosono have all given freely with their time, their encouragement and their brutal honesty! This book would never have been written without their interest in my work. There are many other colleagues and friends too numerous to name who have stimulated me with results, problems and discussions, and I thank them all.

Having begun this project, I would have stopped after the 20th page or so had it not been for my wife, Dora, who has processed all my scribbled notes into a readable manuscript. All the errors within are mine though, and I must apologize in advance to the reader for any slips and mistakes that still persist. It seems that the process of correction is an infinite iterative scheme which does not converge as quickly as one would like.

Dora, Tommy, Chris, and James have been patient while I have spent much spare time jotting notes and examples. These notes would never have been completed without their support.

The major part of this book was drafted while I was at the Mathematical Institute at the University of Oxford, holding an SERC Advanced Fellowship. It was completed after I had joined Intera Sciences, Henley-on-Thames, UK, and I thank my friends there for their interest and encouragement.

Henley-on-Thames P.G.
February 1991

Contents

4. A geometrical theory for waves

5. Nonlinear dispersal mechanisms

6. Flux driven waves in geochemistry

Introduction

This book is concerned with mathematical techniques that can be applied to a large class of nonlinear parabolic partial differential equations. We shall refer to such equations as *reaction-diffusion* systems, even though our sphere of interest will include many cases that are not generally put in this category (e.g. convection-advection-diffusion, as well as nonlinear diffusion processes). The material presented is based on a number of distinct areas of experience. Firstly, I had written lecture notes from courses I taught to graduate students at the University of Dundee in 1985, and at the Mathematical Institute at Oxford in 1988. Secondly, by working in collaboration with members of (and visitors to) the Centre for Mathematical Biology at Oxford, I was made aware of the need to make mathematical techniques available to the mathematician and nonmathematician alike. Also, that it would be useful to provide a stock of examples in order to assist one's intuition in the modelling process. Thirdly, I observed the growth of interest in nonplanar wave-like phenomena and pattern formation. Here the science of observation has surged ahead of the mathematics previously available. Thus the new approaches (e.g. to scroll waves and spiral waves) represent an advance in the understanding and explanation of exotic wave-like phenomena.

The general aims of the book are as follows:

(i) To provide an introduction to the theory of nonlinear parabolic systems suitable for first year graduate students, either as the basis of a course of lectures or as background reading.

(ii) To form a compendium of useful techniques that are applicable to reaction-diffusion systems. The idea is that anyone working in areas modelled by such equations can use the book as an easy reference for ideas and methods.

(iii) To illustrate some nonsmoothing aspects or clustering behaviour within a certain class of models. Such systems have been employed recently to model patchy phenomena (in population biology, for example), and there is a growing interest in their solutions and applications.

(iv) To introduce some of the recent work involving nonplanar wave-like phenomena. In particular, the use of asymptotic analysis to describe the evolution of waves with sharp transition layers.

The idea that the book should be suitable for both mathematicians and nonmathematicians alike has led me to adopt a relaxed style, and I have tried to introduce each new technique or concept by working fully through examples. The generalizations come afterwards, and are sometimes merely outlined and the reader referred elsewhere.

A word about presentation is in order here. The text is interspersed with boxes, indexed alphabetically. These contain short essays on material related to the current section, and broadly fall into one of three categories:

i) Boxes containing background information which may be well-known to most readers. These can easily be skipped in order to follow the developments in the main text (e.g. Box C: Solvability).

ii) Boxes containing technical details which readers may wish to *take on trust* (at least initially), but will be referred to elsewhere in the main text (e.g. Box D: The spectra of some differential operators).

iii) Boxes containing side-issues, which are not strictly part of the main text, but may be of interest to the reader and point out other directions of study (e.g. Box A: The Fokker-Planck equation).

I have tried to restrict the prerequisites assumed for the reader to as few areas as possible. Basically, the reader should have had some experience with the following: calculus, including basic o.d.e.s (first and second order systems, stabilty of rest points, two-point boundary problems for second order equations); linear p.d.e.s (the heat equation!); vector spaces and basic linear operators.

The book is arranged so that the first chapter, Basic considerations, contains a review of the underlying mathematical ideas (with the exception of singular perturbation theory). If a reader is able to follow most of this, then (hopefully!) the rest of the book should follow with few problems.

The book can be thought of as being in three parts. Chapters 1 and 2 provide the preliminaries. Most of the methods introduced here are taken up elsewhere in the book, in applications and worked examples. Chapter 2 is concerned with pattern formation.

Chapters 3 and 4 introduce plane waves (their existence, construction, and applications), and then move on to develop a theory for nonplanar wave-like structures. Many phenomena can be thought of as wave-like (transient or otherwise), and our development is aimed at initial-value problems as well as more intricate patterned solutions (e.g. scroll waves and spiral waves).

Chapter 5 is concerned with nonlinear transport and some nonsmoothing aspects of parabolic systems. Much of this is explained by reference to examples from population biology and elsewhere. We also consider nonlinear diffusion (e.g. porous media, long-range dispersal, etc.) and give some idea of how this affects the results and analysis developed earlier.

The notation used is standard. There is a list of function spaces and other objects included at the back of the book. Wherever possible, I have used subscripts to denote partial differentiation, and also, where no confusion arises, to denote ordinary differentiation. It has the advantage of making the independent variable explicit and of providing continuity (from time-dependent problems to the associated steady-state equations, for example). Where this is likely to cause any misunderstanding, I have reverted to the usual, more formal, notation.

The bibliography is meant to be introductory rather than exhaustive, and I have tried to include textbooks covering some of the material introduced here in order to provide alternative sources.

1 Basic considerations

1.1 Introduction

What are reaction-diffusion equations? Where do they arise? Can we describe their solutions?

The aim of this chapter is to introduce the central objects of our studies. In doing so we shall develop some of the basic mathematical ideas that will be required later on.

This chapter is not meant to comprise a complete review or an in-depth exposition of analytical theory; rather it is a short tour of reaction-diffusion problems and solutions.

Much of the material may be familiar, but the aim here is to present it in the context of reaction-diffusion and restrict the more theoretical considerations to those which impinge on our chosen course of study.

There are many types of reaction-diffusion systems: some tame, some not so tame; some simple, and some which display a richness of subtle and exotic behaviour. In the following chapters we shall meet many interesting and, I hope, exciting problems. But first we must begin.

Perhaps the simplest reaction-diffusion system is the **diffusion equation** itself [41],

$$u_t = \Delta u. \tag{1.1.1}$$

Here $u = u(\mathbf{x}, t)$ is the state variable representing the density or concentration of some entity or substance, at time t and position $\mathbf{x}$ in $\mathbf{R}^n$. Δu denotes the usual Laplacian of u with respect to the space variables $\mathbf{x}$. There is a list of notation and function spaces at the end of the book.

If (1.1.1) holds for all $\mathbf{x} \in \mathbf{R}^n$, say, then once appropriate initial conditions

$$u(\mathbf{x}, 0) = u_0(\mathbf{x})$$

have been specified, it may be solved by Fourier transform methods.

If (1.1.1) holds on some subset Ω of $\mathbf{R}^n$ then we must impose boundary conditions upon u at $\partial\Omega$, the boundary of Ω. For example we might demand that u takes prescribed values on $\partial\Omega$.

The behaviour of solutions of (1.1.1) is well understood and there are many methods (e.g. Fourier transform, eigenfunction expansions – separation of variables [22] [41]) which allow solutions to be represented in terms of *known* quantities.

Notice that (1.1.1) allows us to pose a straightforward question: if we know how u is distributed at time $t = 0$, how will it evolve through time? For this reason it is natural to refer to this and similar systems as **evolution** equations.

Unfortunately (1.1.1) has a simple property which evolution equations generally lack: **linearity** (if $u_1(\mathbf{x},t)$ and $u_2(\mathbf{x},t)$ are solutions then so is $\alpha_1 u_1(\mathbf{x},t) + \alpha_2 u_2(\mathbf{x},t)$ for any constants α_1, α_2). It is this property which makes (1.1.1) *solvable* via transform or expansion techniques. There are many interesting questions to be asked about the solution of linear equations and we stress that there are many difficulties for general initial-boundary-value problems despite their linearity!

Our next example is definitely nonlinear. Consider

$$u_t = \Delta u + f(u). \tag{1.1.2}$$

Here $f(u)$ is some smooth function $:\mathbf{R} \to \mathbf{R}$. We have simply added a **reaction**-term, f, on to our diffusion equation. Again we have $\mathbf{x} \in \Omega$, some open set in $\mathbf{R}^n$, and we shall impose some boundary conditions upon u at $\partial\Omega$.

Suppose

$$f(u) = au + bu^2 + cu^3.$$

It is clear that f, and therefore (1.1.2), is not linear. Because of this there is no hope that any of the methods developed to represent solutions of (1.1.1) will work here. However, even though we may never possess exact solutions of (1.1.2) (in closed form or as expansions), there are many things that we may try to say about (1.1.2) and its solutions. Do they exist? How do they behave? What do they look like?

This is the challenge of nonlinear evolution equations. We must accept that we can hardly ever write down closed form solutions – instead, we must reach for more qualitative analytical techniques which allow us to describe solutions and predict their behaviour.

The equations that will concern us are all broadly of the form

$$u_t = \Delta u + f(u, \nabla u, \mathbf{x}, t).$$

Here the reaction term f is allowed to depend upon the first derivatives of the state variable u (as well as u itself), and may possibly be explicitly dependent upon the independent variables $\mathbf{x}$ and t. Strictly speaking this equation is a *semi-linear parabolic equation* (it is linear in the higher-order derivatives – hence *semi-linear*): we shall continue to refer to such equations as *reaction-diffusion* equations, even though this term tends to be used mainly when f is independent of ∇u.

For example, Burger's equation

$$u_t = u_{xx} - uu_x$$

is included in our sphere of interest, even though the nonlinearity is not thought of as a *reaction*.

Often two or more equations are coupled through the reaction terms forming more complicated systems. The possibilities seem endless ...

Before things become too daunting let us return to the present task. In this chapter we shall begin to confront our nonlinear equations and develop some basic techniques. We begin by seeing how reaction-diffusion equations arise from simple ideas involving conservation laws.

1.2 General balance laws

The reaction-diffusion equations we shall consider arise in many areas of chemistry, biology, ecology, and physics. The equations may be utilized to model a vast variety of phenomena. Why should this be? Well, the use of reaction-diffusion equations often follows from simple time and space dependent models whenever there is a general balance law. We describe their derivation as follows.

When modelling the dispersive behaviour of populations (e.g. of cells or animals) or concentrations (e.g. of chemicals) we often use a continuum approach employing **density functions** to describe the distribution of basic *particles.*

Let $c(\mathbf{x},t) : \Omega \times \mathbf{R}^+ \to \mathbf{R}$, where $\Omega \subseteq \mathbf{R}^n$, be the *particle* density function or concentration. Let $Q(\mathbf{x},t,..)$ be the net creation rate of particles at $\mathbf{x} \in \Omega$ at time t (e.g. the birth rate per unit volume minus the death rate per unit volume). Let $\mathbf{J}(\mathbf{x},t,..)$ be the flux density: that is, for any unit vector $\mathbf{n} \in \mathbf{R}^n$ the scalar product $\mathbf{J}.\mathbf{n}$ is the net rate at which particles cross a unit area in a plane perpendicular to $\mathbf{n}$ (positive in the $\mathbf{n}$ direction).

Now for any regular subset $B \subseteq \Omega$

$$\int_B c\, dx$$

denotes the population mass in B. Here we write dx for the volume increment, $dx_1 dx_2 \ldots dx_n$ when $\mathbf{x} = (x_1, x_2, \ldots, x_n)^T$ in Cartesian coordinates.

We assume that the rate of change of this mass is due to particle creation or degradation inside B, and the inflow and outflow of particles through the boundary ∂B. That is

$$\frac{d}{dt}\int_B c\, dx = -\int_{\partial B} \mathbf{J}.\mathbf{n}\, dA + \int_B Q\, dx$$

where $\mathbf{n}$ denotes the outward-oriented normal to B on ∂B. Assuming the underlying fields are smooth ($\mathbf{J} \in C^1$) we apply the divergence theorem to the first integral on the right and differentiate through the integral on the left to obtain

$$\int_B c_t\, dx = \int_B -\nabla.\mathbf{J} + Q\, dx. \tag{1.2.1}$$

But B was arbitrary in Ω so we must have

$$c_t = -\nabla.\mathbf{J} + Q, \tag{1.2.2}$$

(since if (1.2.2) is violated at some point then we may choose B to be a small n-dimensional ball centred at that point; but then (1.2.1) cannot hold).

This is our balance law. For a given model we must specify Q and **J**. For example, we may follow the theory of diffusion founded by the physiologist Fick. According to **Fick's law** the flux **J** is proportional to the gradient in the density. Thus

$$\mathbf{J} = -D\nabla c.$$

Here the constant D is positive and is called the **diffusivity**. The minus sign indicates that particles are transported from high to low densities. Using Fick's law (1.2.2) becomes a reaction-diffusion equation:

$$c_t = D\Delta c + Q(\mathbf{x}, t, c, ..).$$

There are many more formulations for the flux terms in diffusive processes. Okubo [57] provides a good account of such processes applied in biology. In Box A below we outline an approach based on the random motion of particles.

Example

Two chemicals A and B react according to the rule

$$\mathrm{A} + \mathrm{B} \rightarrow 2\mathrm{B} + \text{other products}.$$

Let a denote the concentration of A and b denote the concentration of B. In a well stirred reaction the concentrations may be assumed to satisfy the **rate reaction** ordinary differential equations:

$$\begin{aligned} a_t &= -\,ab \\ b_t &= ab. \end{aligned}$$

If both chemicals are able to diffuse within the underlying medium the (unstirred) reaction satisfies

$$\begin{aligned} a_t &= D_1\Delta a - ab \\ b_t &= D_2\Delta b + ab. \end{aligned}$$

Box A: The Fokker-Planck equations

We consider a population of individual particles each of which moves according to some stochastic process. In particular we are interested in particles whose individual velocities are random deviations from some externally applied convection velocity. The equation governing such dispersal is a diffusion-type equation (and the process reduces to Fickian diffusion in the absence of convection).

For simplicity we shall restrict our attention to one dimensional dispersal, so our random variables will live in $\mathbf{R}$. The generalisation to n-dimensions follows in an analogous manner [66].

We first define the conditional density of the random variable $\mathbf{x}$ to be the limit

$$P(x,t,y,s) = \lim_{h\to 0} \frac{\text{Prob.}\{x \leq \mathbf{x}(t) \leq x+h : \text{when } \mathbf{x}(s) = y\}}{h}$$

(here $t \geq s$). In other words, for δx small, $P(x,t,y,s)\delta x$ is the probability that $\mathbf{x}$ is in $[x, x+\delta x]$ at time t, given that it started out at y at time s.

Now suppose that a population of particles are each located by such random variables. Let $u(x,t)$ denote the density distribution of the population in $\mathbf{R}$, at time t. It is clear that u should satisfy

$$u(x,t) = \int_{\mathbf{R}} P(x,t,y,s)u(y,s)\, dy, \tag{1}$$

for $t > s$. Notice we have $P(x,t,y,s) \to \delta(x-y)$ as $t \to s$ ($\delta(.)$ denotes the Dirac point mass distibution, see [22] for example).

Now using (1) we shall derive a differential equation that u must satisfy. To do this we shall utilize the behaviour of P, as embodied in the following quantities known as moments:

$$M_k(t,y,s) = \int_{\mathbf{R}} P(z,t,y,s)(z-y)^k\, dz. \tag{2}$$

Now since we have assumed P to be well defined and (hopefully) bounded for $t > s$, we have

$$P(x,t,y,s) = \int_{\mathbf{R}} \delta(z-x)P(z,t,y,s)\, dz. \tag{3}$$

Using the formal Taylor series expansion for δ, that is

$$\begin{aligned}\delta(z-x) &= \delta(y-x+z-y)\\ &= \sum_{k=0}^{+\infty} \frac{(z-y)^k}{k!}\left(\frac{\partial}{\partial y}\right)^k \delta(y-x)\\ &= \sum_{k=0}^{+\infty} \frac{(z-y)^k}{k!}\left(-\frac{\partial}{\partial x}\right)^k \delta(y-x),\end{aligned}$$

we substitute into (3) to obtain;

$$\begin{aligned}P(x,t,y,s) &= \sum_{k=0}^{+\infty} \frac{1}{k!}\left(-\frac{\partial}{\partial x}\right)^k \int_{\mathbf{R}} (z-y)^k P(z,t,y,s)\,dz\,\delta(y-x)\\ &= \left(1+\sum_{k=1}^{+\infty}\frac{1}{k!}\left(\frac{-\partial}{\partial x}\right)^k M_k(t,y,s)\right)\delta(y-x)\\ &= \left(1+\sum_{k=1}^{+\infty}\frac{1}{k!}\left(\frac{-\partial}{\partial x}\right)^k M_k(t,x,s)\right)\delta(x-y). \qquad (4)\end{aligned}$$

Here we have used (2) together with the simple rules $\delta(y-x) = \delta(x-y)$ and $\delta(y-x)h(y) = \delta(y-x)h(x)$ (for bounded functions h).

Now substituting (4) into (1) we obtain:

$$\begin{aligned}u(x,t)-u(x,s) &= \sum_{k=1}^{\infty}\left(-\frac{\partial}{\partial x}\right)^k \int_{\mathbf{R}} \delta(x-y)\frac{M_k(t,x,s)}{k!}u(y,s)dy\\ &= \sum_{k=1}^{\infty}\left(-\frac{\partial}{\partial x}\right)^k \frac{M_k(t,x,s)}{k!}u(x,s).\end{aligned}$$

Dividing by $(t-s)$ and letting $t \to s$ we obtain:

$$u_t(x,t) = \sum_{k=1}^{\infty}\left(-\frac{\partial}{\partial x}\right)^k (K_k(x,t)u(x,t)) \qquad (5)$$

where the $K_k(x,t)$ are the Kramer coefficients,

$$K_k(x,t) = \lim_{h\to 0}\frac{M_k(t+h,x,t)}{h}. \qquad (6)$$

Now (5) is our partial differential equation describing the evolution of the density function u. Clearly $P(x,t,y,s)$ is a (fundamental) solution of (5).

However, the coefficients are themselves defined via P, so we do not appear to have got very far! Fortunately it is possible to evaluate the Kramer coefficients by considering the underlying random process governing the motion of each particle. Having done this we can then solve (5) for P if we wish.

There are many ways to proceed, depending upon the choice of stochastic process governing particle motion. Risken's book [66] contains a good account of many refinements. However, common to all processes of interest is the fact that all the coefficients in (6) vanish for $k > 2$. Thus, (5) is a second-order equation and is known as the Fokker-Planck equation, or Kolmogoroff's forward equation. That is

$$u_t = (K_2(x,t)u)_{xx} - (K_1(x,t)u)_x.$$

For simplicity we consider the following process:

$$\frac{dx}{dt} = c(x,t) + \frac{dw}{dt}$$

where $w(t)$ is a Wiener-Lévy process [62] with independent increments: that is $w(t)$ is normally distributed with mean $E(w(t)) = 0$ and $w(0) = 0$. More generally it has the expected values $E(w(t) - w(t_0)) = 0$ and $E((w(t) - w(t_0))^2) = \alpha^2(t - t_0)$, for some constant α. In fact we can write $w(t) = \int_0^t n(s)ds$ where n is a stationary normal process with zero mean and flat spectrum (white noise). The term $c(x,t)$ represents the average velocity of individuals at x at time t, and is assumed to be determined by some process outside of our current sphere of interest.

Now

$$dx = c(x,t)dt + dw$$

so

$$E(dx) = c(x,t)dt.$$

Moreover,

$$\begin{aligned} E((dx)^2) =& c^2(x,t)dt^2 + 2c(x,t)dtE(dw) + E(dw^2) \\ =& c^2(x,t)dt^2 + \alpha^2 dt. \end{aligned}$$

But these are precisely the first two moments, $M_1(t+dt,x,t) = E(dx)$ and $M_2(t+dt,x,t) = E(dx^2)$. Thus $K_1(x,t) = c(x,t)$ and $K_2(x,t) = \alpha^2$. So u satisfies

$$u_t = \alpha^2 u_{xx} - (uc(x,t))_x,$$

a one dimensional diffusion equation. The convective velocity $c(x,t)$ is common to all individuals in the population and must be specified in a given situation.

For example, the process of **chemotaxis** in biology is one in which individual mobile cells are assumed to move up gradients of a particular chemical concentration (known as a chemo-attractant). In cases such as **slime moulds**, the cells are themselves secreting the chemo-attractant as a means of forming local clusters. Assuming the attractant concentration, denoted by $v(x,t)$ is subject to random, unbiased diffusion and linear degradation we have a simple one-dimensional model:

$$\begin{aligned} u_t &= \alpha^2 u_{xx} - (uv_x)_x \\ v_t &= Dv_{xx} - rv + s(u). \end{aligned}$$

Here we have set $c(x,t) = v_x(x,t)$, so that individuals have random motion biased by the gradient of the chemo-attractant. The term $s(u)$ represents the (nonlinear) secretion of attractant by the cells, and D is the diffusivity.

We shall look more closely at such equations in Chapter 5.

Now consider the equation

$$\begin{gathered} u_t = \Delta u + f(u), \\ \mathbf{x} \in \Omega \subseteq \mathbf{R}^n,\ t > 0, \end{gathered}$$

together with initial conditions

$$u(\mathbf{x}, 0) = u_0(\mathbf{x})$$

some given function on Ω. If $\Omega \neq \mathbf{R}^n$ then we specify some **boundary conditions** at $\partial\Omega$. These have the form

$$B(\mathbf{x}, t, u, \nabla u) = 0, \quad \mathbf{x} \in \partial\Omega,\ t > 0.$$

For example:

(i) **Neumann** or **no-flux** boundary conditions:

$$\nabla u.\mathbf{n} = 0, \quad \mathbf{x} \in \partial\Omega,\ t > 0.$$

Here $\mathbf{n}$ is the outer normal to Ω at $\mathbf{x} \in \partial\Omega$. For equations like the one above this means that there can be no flux of particles either into or out of the domain Ω.

(ii) **Dirichlet** boundary conditions:

$$u = b(\mathbf{x}, t), \quad \mathbf{x} \in \partial\Omega,\ t > 0.$$

Here b is some prescribed function. The conditions are called **homogeneous** if $b \equiv 0$.

(iii) **Robin** or **mixed** boundary conditions:

$$\alpha(\mathbf{x},t)u + \beta(\mathbf{x},t)\nabla u.\mathbf{n} = b(\mathbf{x},t), \quad \mathbf{x} \in \partial\Omega, \ t > 0.$$

Here $\mathbf{n}$ and b are as in (i) and (ii) and α and β are given functions.

We might also impose combinations of (i)-(iii) on separate parts of the boundary. These conditions are all linear in the state variable u, but we could just as easily apply nonlinear ones (although this may well make the analysis a lot harder).

If boundary conditions are required, then their determination is really part of the modelling process. The behaviour of solutions will certainly be affected by the choice of conditions. Hence, as much care should be taken in modelling the correct boundary conditions as is taken in modelling the evolution equation itself.

In section 1.6 we will see how the theory of the existence of solutions to reaction-diffusion equations relies on properly posed boundary conditions and the choice of *reasonable* initial data $u_0(\mathbf{x})$. For the moment we will pass over these questions and concentrate instead upon some of the basic properties of reaction-diffusion problems and their solutions.

Box B: Variational derivatives and gradient systems

Before leaving the present section concerned with how reaction-diffusion equations arise, we shall make a few remarks about some special systems which may be derived via (or at least considered to be due to) **variational** considerations.

Let p be a real-valued function defined on some dense subspace of $L_2(\Omega)$. The **variational derivative** of p is defined as follows. For u fixed in the domain of p, we define the linear functional G_u by

$$G_u(v) = \lim_{h\to 0} \frac{p(u+hv) - p(u)}{h} .$$

Here G_u is initially defined for smooth functions v such that $u + hv$ lies in the domain of p, and then extended to have as large a domain as possible. In particular, if p is reasonably well behaved and u is smooth enough, then G_u may be extended to be a bounded linear functional on $L_2(\Omega)$ (and thus be defined on the whole space). We shall assume this to be the case.

Now, by the Riesz representation theorem [42], there exists an element w_u in $L_2(\Omega)$ such that

$$G_u(v) = < v, w_u >$$

(where $< v, w >$ denotes the usual L_2 inner product, $\int_\Omega vw^* dx$, where * denotes complex conjugation).

Now we define the variational derivative of p at u to be w_u. Clearly for any v in $L_2(\Omega)$ with $\|v\| = 1$, $< v, w_u >$ defines the directional derivative of p at u, in the v direction.

(The analogy of the finite-dimensional vector space $\mathbf{R}^n$ is of interest here. If p: $\mathbf{R}^n \to \mathbf{R}$ then $w_u \in \mathbf{R}^n$ may be defined as above and is recognizable as the gradient of p at u, denoted ∇p.)

We will denote the variational derivative w_u by grad(p) or ∇p.

Now let us suppose that $u \in L_2$ denotes the *state* of some physical or chemical system and that there is a generalized *energy* defined for each state, of the form

$$E(u) = \int_\Omega F(u, \nabla u, \mathbf{x})\, dx.$$

We suppose further that u satisfies a no-flux boundary condition $\nabla u.\mathbf{n} = 0$ on $\partial\Omega$ (it is relatively easy to incorporate Dirichlet or other conditions). This condition would be incorporated implicitly into the definition of the domain of E.

The requirement that $E < \infty$ imposes a natural integrability (smoothness) condition upon u, so we will assume that the domain of E is dense in L_2.

To introduce a dynamic element we assume that the state u evolves through time in such a way as to decrease E continually. We set

$$u_t = -\nabla E(u) \tag{1}$$

so that

$$\frac{dE}{dt} = -\|\nabla E\|^2.$$

The definition (1) characterizes a class of equations known as **gradient systems.**

For example, if

$$E(u) = \int_\Omega \frac{|\nabla u|^2}{2} + H(u)\, dx$$

then

$$E(u + hv) - E(u) = h\left(\int_\Omega (\nabla u.\nabla v + \frac{dH}{du}v)\, dx\right) + O(h^2).$$

Thus, integration by parts yields

$$G_u v = \int_\Omega (\nabla u.\nabla v + \frac{dH}{du})dx = \int_\Omega (-\Delta u + \frac{dH}{du})v\, dx + \int_{\partial\Omega} v\nabla u.\mathbf{n}\, dA.$$

The second term on the right is annihilated by the boundary condition (note that v also satisfies a no-flux boundary condition). Initially for G_u to exist we require ∇v to be in L_2, whereas after the integration by parts we only need v in L_2. The price we pay is the extra smoothness that must be assumed for u (i.e. $\Delta u \in L_2$).

Hence, $\nabla E(u) = -\Delta u + \frac{dH}{du}(u)$. So our evolution equation becomes

$$u_t = \Delta u + \frac{dH}{du}(u).$$

For more general systems we may have a constraint of the form

$$\int_\Omega g_1(u, \nabla u, \mathbf{x})\, dx = 0.$$

In this case we augment E to become

$$\int_\Omega f(u, \nabla u, \mathbf{x}) + \lambda g_1(u, \nabla u, \mathbf{x})\, dx.$$

Here λ is independent of $\mathbf{x}$ and plays the role of a Lagrange multiplier. We get an evolution equation involving λ, which is allowed to be a function of time and assumes values so as to maintain the constraint.

If we have a constraint of the form

$$g_2(u, \nabla u, \mathbf{x}) = 0, \quad \mathbf{x} \in \Omega,$$

then we allow the multiplier λ in the augmented energy to depend on $\mathbf{x}$.

All of the above may be extended to cope with functionals defined on two or more state variables. We illustrate this with an example.

Example

Suppose $\mathbf{u} = (u, v)$: $\mathbf{R}^2 \to \mathbf{R}^2$ denotes a pair of state variables defined on (x, y) space. Consider

$$E(u,v) = \int_{Re^2} \frac{(u_x^2 + u_y^2 + v_x^2 + v_y^2)}{2}\, dxdy = \int_{\mathbf{R}^2} \frac{|\nabla \mathbf{u}|^2}{2}\, dxdy$$

subject to the constraint $u_x + v_y = \nabla.\mathbf{u} = 0$.

Then we augment E to become

$$\int_\Omega \frac{|\nabla \mathbf{u}|^2}{2} + \lambda(x,y)\nabla.\mathbf{u}\, dxdy.$$

The associated gradient system is

$$u_t = \Delta u + \lambda_x$$
$$v_t = \Delta v + \lambda_y,$$

or in vector notation,

$$\mathbf{u}_t = \Delta \mathbf{u} + \nabla \lambda,$$

where the *pressure*-like variable λ is always such that $\nabla.\mathbf{u} = 0$.

1.3 A scalar equation

Here we consider a problem involving one state variable u, in one space dimension:

$$\begin{aligned} u_t =& u_{xx} + u(1-u)(u-a), \\ & x \in (0,10), t > 0; \\ u(x,0) =& u_0(x), \quad x \in (0,10); \\ u =& 0, \quad x = 0,\ 10, \quad t > 0. \end{aligned} \tag{1.3.1}$$

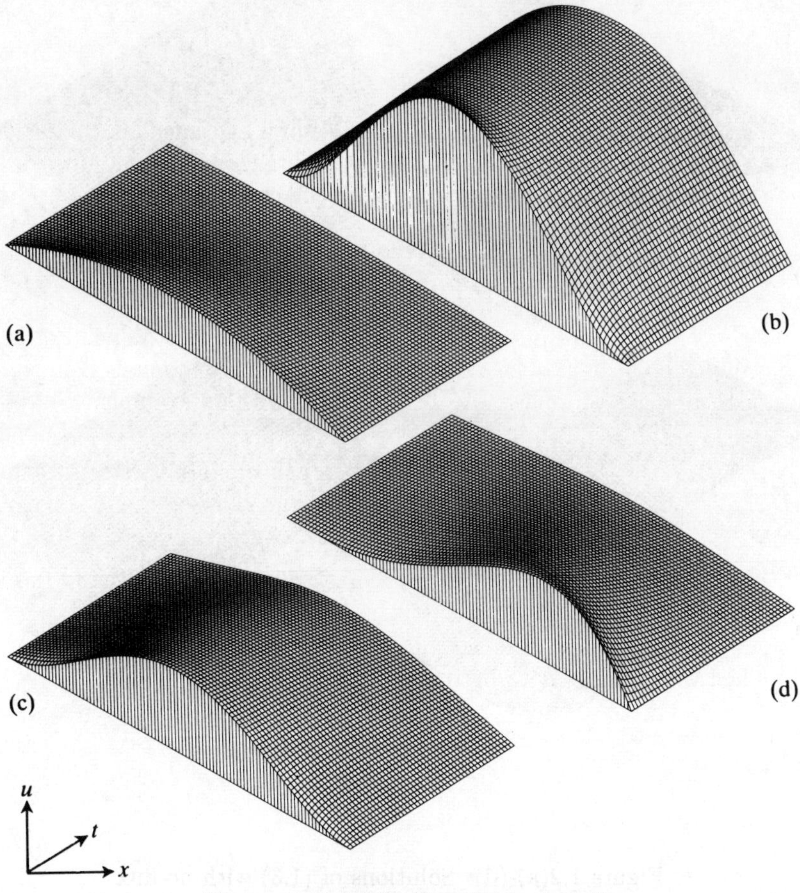

Figure 1.1(a)-(d): Solutions of (1.3).

Here a is a constant in (0,0.5).

Figures 1.1(a)-(d) depict some numerical solutions of the above problem for a variety of initial data $u_0(x)$. Notice that as $t \to \infty$, u either approaches zero or some positive, nonconstant equilibrium. How can we predict which will occur?

We also numerically solved the Neumann problem for the equation in (1.3.1) with $u_x = 0$ at $x = 0$ and 10. This time the solutions approach either $u = 1$ or $u = 0$ as $t \to \infty$. It is easy to see that both are equilibria for the Neumann problem (so is $u = a$, but we found no solutions approaching this as $t \to \infty$: why? See the next section!).

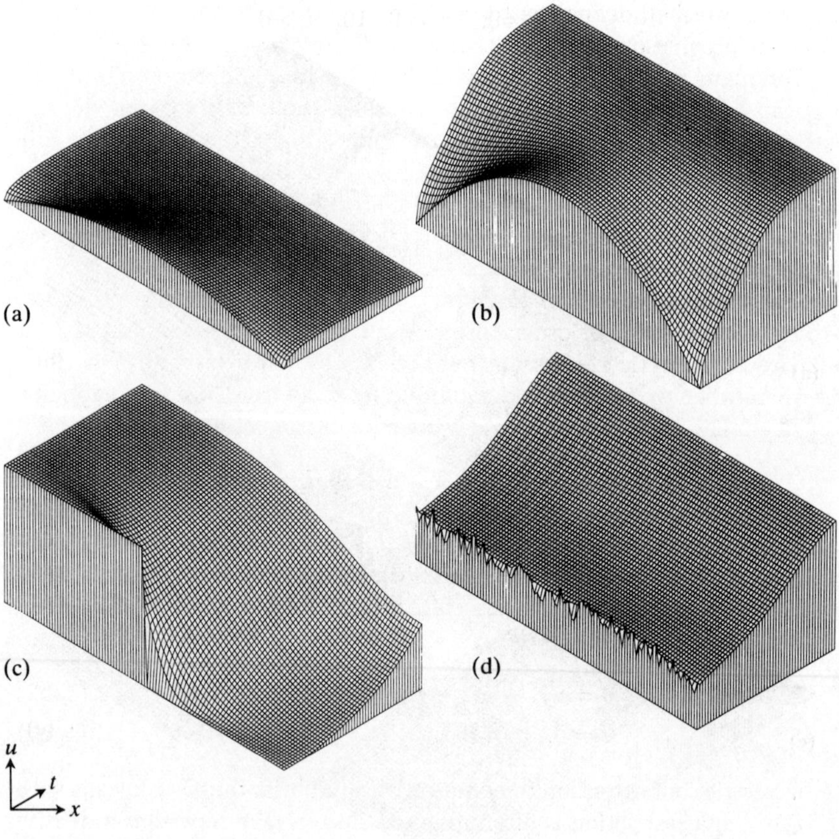

Figure 1.2(a)-(d): Solutions of (1.3) with no-flux boundary conditions.

1.4 Equilibria and linear stability

Given a reaction-diffusion system it is often important to analyse the possible long-term behaviour of solutions. In particular, we may wish to know whether the solutions will settle down to some time-independent **equilibrium**. Of course, this need not necessarily be the case. The long-time behaviour may be periodic, aperiodic, or chaotic.

We refer to any such long-time regimes as **attracting** if they are able ultimately to dominate the behaviour of solutions which start out nearby.

The simplest kind of **attractors** are equilibria which are stable to small perturbations. Thus, for any given problem, it is important to find and classify the possible equilibria according to their stability properties.

We proceed by considering some examples below. After locating some equilibria, we shall define the term *linear stability* and show how one may set about proving qualitatively that an equilibrium is linearly stable.

For many nonlinear systems the location of equilibria is a problem that can only be approached qualitatively. If parameters such as domain size or reaction rates can be varied, we may discuss the appearance and disappearance of equilibria via the techniques of bifurcation theory. This is a vast subject in its own right, so we shall borrow just as much as we need and try to focus on our specific examples. A more general approach may be found in [23], [8], and Box G in Chapter 2.

In Chapter 2 we shall employ some other ideas (such as those of singular perturbation theory) in order to construct equilibria. The point is that bifurcation theoretic results are *local* results by nature, so we must look elsewhere when we want to find equilibria far away from bifurcation points.

The following problem provides a simple introduction:

$$\begin{aligned} u_t &= u_{xx} + u(1-u), \quad x \in (0, L),\ t > 0, \\ u &= 0, \quad x = 0, L \quad t > 0. \end{aligned} \tag{1.4.1}$$

Here $L > 0$ is the domain length which we may vary. The **steady-state** equation for equilibria associated with (1.4.1) is simply

$$0 = u_{xx} + u(1-u), \quad x \in (0, L), \tag{1.4.2}$$

$$u = 0, \quad x = 0, L. \tag{1.4.3}$$

Let us restrict our attention to nonnegative equilibria, that is solutions with $u \geq 0$. Such a restriction is often required since u may represent a density function or concentration which has to be nonnegative by definition.

Clearly $u \equiv 0$ is an equilibria for any choice of L: are there any more? In this case we can give an answer by using **phase plane** analysis. We sketch solutions of (1.4.2) in the (u, u_x)-plane, or phase plane.

Multiplying (1.4.2) by u_x and integrating with respect to x, we have

$$\frac{u_x^2}{2} + \int^x u(1-u)u_x\,dx = \text{constant}$$

that is,

$$\frac{u_x^2}{2} + \int^u w(1-w)dw = \text{constant}$$

$$\frac{u_x^2}{2} + \frac{u^2}{2} - \frac{u^3}{3} \equiv H(u, u_x) = \text{constant}.$$

Thus we may draw the solutions of (1.4.2) as the contours of H: see Figure 1.3.

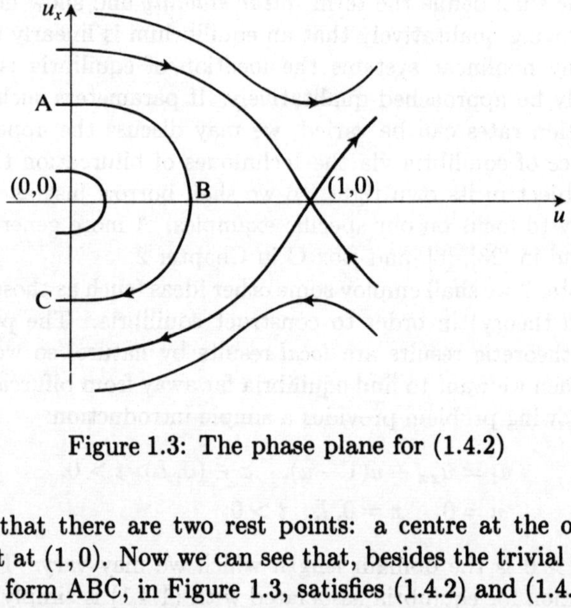

Figure 1.3: The phase plane for (1.4.2)

Notice that there are two rest points: a centre at the origin and a saddle point at $(1, 0)$. Now we can see that, besides the trivial solution, an orbit of the form ABC, in Figure 1.3, satisfies (1.4.2) and (1.4.3) provided that we have $x = 0$ at A and $x = L$ at C. Clearly there is some symmetry about the phase plane (H is an even function of u_x),and so the *time* taken in moving from A to B along the orbit is equal to that taken between B and C.

Let A be the point $(0, p)$ (so C is the point $(0, -p)$), then at A we have $H = \frac{p^2}{2}$, while at $(1, 0)$, $H = \frac{1}{6}$. Thus, we must have $0 < p < \frac{1}{\sqrt{3}}$, since if p is too large the orbit through A will never meet the u-axis.

Now, on the orbit through A we have

$$|u_x| = \sqrt{\frac{2}{3}u^3 - u^2 + p^2},$$

so between A and B, where u_x is nonnegative, we may integrate to obtain,

$$x_B = \int_0^{u_B} \frac{du}{\sqrt{p^2 - u^2 + \frac{2}{3}u^3}},$$

where $u(0) = 0$, and x_B and u_B are the values of x and u at B. In fact u_B is the positive zero of $p^2 - u^2 + \frac{2}{3}u^3$ (which exists and lies in (0,1) so long as $0 < p < \frac{1}{\sqrt{3}}$).

Thus, if $x_B = L/2$, we have a nonconstant solution of the boundary value problem (1.4.2)-(1.4.3). Rather than fixing L and trying to find a corresponding value of u_B (and hence p), it is easier to express L as a function of u_B as it varies in (0,1). We have

$$L = 2\int_0^{u_B} \frac{du}{\sqrt{u_B^2 - u^2 + \frac{2}{3}(u^3 - u_B^3)}}. \tag{1.4.4}$$

We call this a **time-map** since it relates the *time* taken to traverse an orbit segment, to a point (in this case $(u_B, 0)$) on that segment. The idea of time-maps often proves useful in second-order steady-state problems (see [70],[71],[72]).

The expression (1.4.4) is an elliptic integral, but we may evaluate it numerically to obtain a graph of the equilibria as L varies. Such a graph is called a **bifurcation diagram**. Figure 1.4 depicts the current situation. Here $\|u\|_\infty = \max\{|u| : x \in [0,1]\}$ is equal to u_B for the nonconstant solutions.

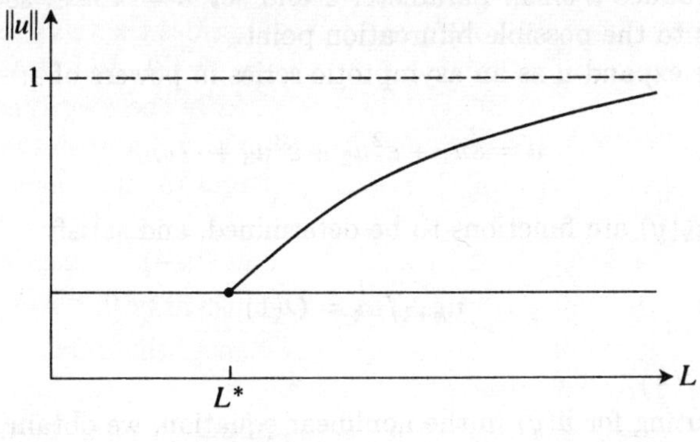

Figure 1.4: Bifurcation diagram for (1.4.2)-(1.4.3)

Notice that the nontrivial solutions branch away, or **bifurcate**, from the trivial solutions at $L = L^*$, which we may calculate from (1.4.4) as

$$L^* = 2 \lim_{\delta \to 0} \int_0^\delta \frac{du}{\sqrt{\delta^2 - u^2}} = \pi.$$

This is an example of **bifurcation**: the point $u = 0, L = \pi$ is called a **bifurcation point**. It is fortunate that in this example we have been able to construct the global bifurcation diagram (i.e. all the nontrivial solutions). We cannot do this in general, so it is important to have some means of investigating bifurcation points when we cannot obtain any expressions like (1.4.4). Such techniques are limited to describing the behaviour of the bifurcation curve in a neighbourhood of the bifurcation point, but are extremely useful in explaining the development of structure in reaction-diffusion systems (see Chapter 2).

We illustrate these ideas by continuing with the present example, concentrating on the behaviour near the bifurcation point. Let us imagine that we have not seen Figure 1.4 and begin again from (1.4.2)-(1.4.3).

We set $y = x/L$, so our problem becomes

$$u_{yy} + L^2 u(1-u) = 0 \quad y \in (0,1); \quad u = 0, \quad y = 0, 1.$$

This fixes the domain and introduces the parameter L into the equation.

Now close to the trivial solution $u = 0$, we can approximate the above nonlinear equation by linearizing. The resulting equation $u_{yy} + L^2 u = 0$, together with the boundary conditions, $u = 0$ at $y = 0$ and 1, has nontrivial, nonnegative solutions if and only if $L = \pi$ (when $u = (\text{constant}).\sin \pi y$). This suggests that nonconstant solutions of the nonlinear problem may bifurcate from $(L, u) = (\pi, 0)$.

We introduce a small parameter ε and set $L = \pi + \varepsilon$, so that $\varepsilon = 0$ corresponds to the possible bifurcation point.

Now we expand u as an asymptotic series in powers of ε:

$$u = \varepsilon u_1 + \varepsilon^2 u_2 + \varepsilon^3 u_3 + \cdots,$$

where the $u_k(y)$ are functions to be determined, and satisfy

$$u_{k+1}/u_k = O(1),$$

(i.e. never $O(\frac{1}{\varepsilon})$).

Substituting for $u(y)$ in the nonlinear equation, we obtain

$$\varepsilon(u_{1\,yy} + \pi^2 u_1) + \varepsilon^2(u_{2\,yy} + \pi^2 u_2 + 2\pi u_1 - \pi^2 u_1^2) + O(\varepsilon^3) = 0.$$

Now we determine the functions $u_k(y)$ by equating the coefficients of successive powers of ε to zero in ascending order.

To order ε we have

$$u_{1\ yy} + \pi^2 u_1 = 0, \quad u_1(0) = u_1(1) = 0.$$

This is just the linearization of the original equation about $u = 0$ with $L = \pi$, and thus has the solution

$$u_1 = A \sin \pi y,$$

where A is some constant still to be determined.

To order ε^2 we have

$$u_{2\ yy} + \pi^2 u_2 = \pi^2 u_1^2 - 2\pi u_1, \quad u_2(0) = u_2(1) = 0.$$

Now multiplying this equation by u_1 and integating over [0,1] we see that

$$0 = \int_0^1 (\pi^2 u_1^2 - 2\pi u_1) u_1 \, dx.$$

Conversely, if this last expression is satisfied, then we are able to solve the equation for u_2 (see Box C). Thus we have a **solvability** condition that must be satisfied by u_1. We use it to determine the constant A. We obtain

$$A\pi \int_0^1 \sin^3 \pi y \, dy = 2 \int_0^1 \sin^2 \pi y \, dy.$$

Thus, $A = 3/4$.

Having determined u_1 exactly we can now go on to determine u_2 from the inhomogeneous equation if we wish. We leave this as an exercise.

We have obtained the asymptotic solution

$$u = \frac{3}{4}(L - \pi) \sin \pi y + O((L - \pi)^2),$$

which is nonnegative for $L \geq \pi$. Hence, we have captured the local behaviour at the bifurcation point in Figure 1.4.

The above asymptotic technique was because we chose the correct form for the expansion of u, (that is integer powers of ε). Suppose instead we had expanded u in powers of ε^α:

$$u = \varepsilon^\alpha u_1 + \varepsilon^{2\alpha} u_2 + \cdots,$$

where $u_1 \not\equiv 0$. Then we would obtain $u_1 = A \sin \pi y$, as before; while the next highest terms would be

$$\varepsilon^{2\alpha}(u_{2\,yy} + \pi^2 u_2 - \pi^2 u_1^2) + \varepsilon^{\alpha+1} 2\pi u_1.$$

Now, if $\alpha > 1$, then $2\alpha > \alpha + 1$, so equating the coefficient of the smaller power of ε to zero would yield $u_1 = 0$ – a contradiction. If $\alpha < 1$ then 2α would be the smaller power, and we would have

$$u_{2\,yy} + \pi^2 u_2 = \pi^2 u_1^2, \quad u_2(0) = u_2(1) = 0.$$

Applying the solvability criterion, as above, we would see that this last equation could have no solution unless $u_1 = 0$ – again a contradiction. Hence, the choice of $\alpha = 1$ in the above expansion for u would have been forced upon us had we not made our fortunate *guess*.

A more general approach to bifurcation, known as **Liapunov-Schmidt** theory, is discussed in Box G, in Chapter 2. This is more rigorous, but essentially contains the same ingredients as the above example.

Box C: Solvability

Consider

$$v_{xx} + q(x)v = h(x), \quad x \in (0,1), \quad v(0) = v(1) = 0. \tag{1}$$

Here q and h are some given functions.

Suppose $w(x)$ is a solution of the associated homogeneous problem:

$$w_{xx} + q(x)w = 0, \qquad x \in (0,1), \quad w(0) = w(1) = 0.$$

Multiplying (1) by w and integrating over [0,1], we have

$$\int_0^1 v(w_{xx} + qw)\, dx = \int_0^1 hw\, dx.$$

Thus

$$0 = \int_0^1 h(x)w(x)\, dx \tag{2}$$

is a necessary condition for (1) to have a solution. This is a **solvability** condition. Conversely it can be shown that if (2) holds, then (1) has a unique solution v such that $\int_0^1 vw\, dx = 0$ (construct the Green's function! [5]). For example, if $q = (k\pi)^2$ then

$$v(x) = \sin k\pi x \int_0^x \frac{h(s)\cos k\pi s}{k\pi} ds - \cos k\pi x \int_0^x \frac{h(s)\sin k\pi s}{k\pi} ds.$$

More generally if H is some Hilbert space (a complete vector space whose norm is generated by an inner product), and A is any self-adjoint operator : $H \to H$ (i.e. $< Ax, y >=< x, Ay >$ for all $x, y \in D((A)$,the domain of A), then for b given in H,

$$Ax = b$$

has a solution x only when b is orthogonal to every vector in the null space of A (i.e. vectors y such that $Ay = 0$). This is a solvability criterion, and, if it is satisfied, the solution x is uniquely determined up to the addition of any vectors in the null space of A. We refer to Box G, in Chapter 2, and Kreyzig's book [42] for a discussion of solvability conditions in this and more general settings where they are usually referred to as the **Fredholm alternative.**

Let us consider a further example, this time with no-flux boundary conditions:

$$u_t = u_{xx} + u(1-u)(u-a), \quad x \in (0, L), \quad u_x(0) = u_x(L) = 0.$$

Here a is a constant in (0,1/2).

Again we consider the associated steady-state problem and rescale the independent variable by setting $y = x/L$. We have

$$u_{yy} + L^2 u(1-u)(u-a) = 0, \quad y \in (0, 1), \quad u_y(0) = u_y(1) = 0. \quad (1.4.5)$$

Trivially $u \equiv 0, a$ and 1 are solutions, while phase plane analysis yields the situation depicted in Figure 1.5.

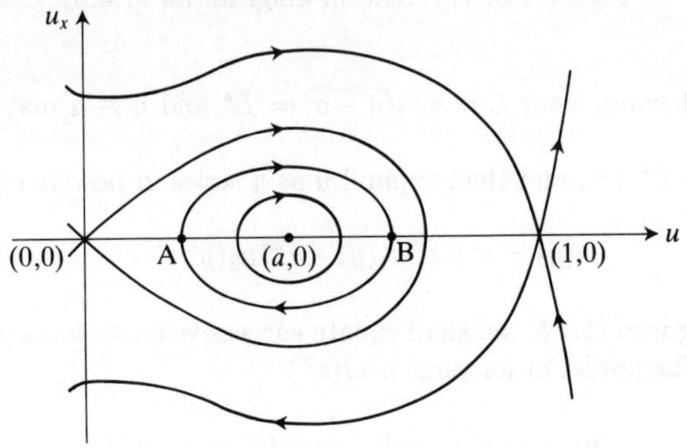

Figure 1.5: The phase plane for (1.4.5).

Clearly any orbit segment of the form AB, BA, ABA, BAB, etc, are candidates for solutions of (1.4.5). Proceeding in the same way as the previous example we see that

$$H(u, u_y) = \frac{u_y^2}{2} + L^2 \int_0^u w(1-w)(w-a)\, dw$$

is constant along orbits. Again we may find the time-map, and compute the bifurcation diagram. In this example, we shall restrict ourselves to considering monotone increasing solutions only (i.e. orbit segments of the type AB). Other solutions can be constructed from these by symmetric continuation to domains of the form $(0, m)$ (m an integer), and then rescaling y.

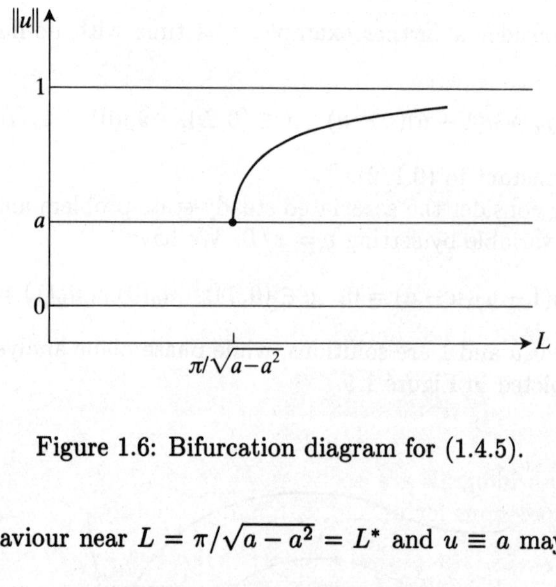

Figure 1.6: Bifurcation diagram for (1.4.5).

The behaviour near $L = \pi/\sqrt{a - a^2} = L^*$ and $u \equiv a$ may be seen as follows.

Set $L = L^* + \varepsilon$, and then expand u as a series in powers of ε^α :

$$u(y) = a + \varepsilon^\alpha u_1(y) + \varepsilon^{2\alpha} u_2(y) + \cdots.$$

Substituting into (1.4.5) we shall equate successive coefficients of powers of ε to zero. The lowest order term is $O(\varepsilon^\alpha)$:

$$u_{1\ yy} + \pi^2 u_1 = 0, \quad u_{1\ y}(0) = u_{1\ y}(1),$$

so that $u_1 = A \cos \pi y$, for some constant A.

The next highest-order term is either $O(\varepsilon^{2\alpha})$:

$$u_{2\ yy} + \pi^2 u_2 = -\pi^2 \frac{(1-2a)}{(a-a^2)} u_1^2, \quad u_{2\ y}(0) = u_{2\ y}(1) = 0; \tag{1.4.6}$$

or else $O(\varepsilon^{1+\alpha})$:

$$2\pi\sqrt{a-a^2}u_1 = 0. \tag{1.4.7}$$

Clearly, since u_1 is to be nonzero, we must have $2\alpha \leq 1 + \alpha$. Moreover if $\alpha = 1$, then both terms are of the same order, and we must include the linear term from (1.4.7) in the right-hand side of (1.4.6). However, the solvability of (1.4.6) requires that the right-hand side is orthogonal to $\cos \pi y$ (see Box C), so we can only allow even powers of u_1 to appear there. Thus, we must have $\alpha < 1$, and we can solve (1.4.6) above for u_2:

$$u_2 = \frac{(1-2a)A^2}{3(a-a^2)}(\cos^2 \pi y - 2) + B \cos \pi y,$$

where B is a free constant still to be determined.

The next highest order term is either (again) $O(\varepsilon^{1+\alpha})$, that is (1.4.7), or else $O(\varepsilon^{3\alpha})$:

$$u_{3\ yy} + \pi^2 u_3 = \frac{\pi^2}{(a-a^2)}(u_1^3 - 2(1-2a)u_1u_2), \quad u_{3\ y}(0) = u_{3\ y}(1) = 0. \tag{1.4.8}$$

Clearly we must have $\alpha + 1 \geq 3\alpha$. If this inequality is strict, then the right-hand side of (1.4.8) must be orthogonal to $u_1(y)$, ie;

$$\int_0^1 u_1^4\, dy + 2(1-2a)\int_0^1 -u_1^2 u_2\, dy = 0.$$

However, both integrals are positive: we can calculate the second one by using the expressions for u_1 and u_2 found previously:

$$2(1-2a)\int_0^1 -u_1^2 u_2\, dy = \frac{5(1-2a)^2 A^4}{12(a-a^2)}.$$

Thus, we must choose $1 + \alpha = 3\alpha$: that is, $\alpha = 1/2$.

Now, including the linear term from (1.4.7) on the right-hand side of (1.4.8), the solvability condition requires

$$\frac{\pi^2}{(a-a^2)}\int_0^1 u_1^4 dy - \frac{2\pi^2(1-2a)}{(a-a^2)}\int_0^1 u_1^2 u_2\, dy = 2\pi\sqrt{a-a^2}.\int_0^1 u_1^2 dy. \tag{1.4.9}$$

Here $u_1 = A\cos\pi y$, while u_2 solves (1.4.6). When (1.4.9) is simplified it yields a quartic for A of the form

$$0 = A^2(A^2 - A_0^2).$$

The constant A_0 can be calculated explicitly, and this is left to the reader (note that the free constant in the solution of (1.4.6) plays no role since the integrals of odd powers of $\cos\pi y$ vanish).

The nonconstant solutions in Figure 1.6 is therefore locally quadratic, near to the bifurcation point. That is,

$$\|u\|_\infty = a + (L - L^*)^{\frac{1}{2}} A_0 + O(L - L^*).$$

Clearly, we could have saved ourselves a lot of work had we guessed $\alpha = 1/2$, and hence the quadratic behaviour of the bifurcation. With practice we may all make such jumps, but it is as well to have some procedure to follow when clarity or inspiration is lacking.

Having considered some examples, we shall defer further discussion of bifurcation theory until Chapter 2, where a more general approach will be outlined. In particular, we shall see how bifurcation is intimately related to the question of **stability**, which we shall consider for the remainder of this section.

An equilibrium solution of a reaction-diffusion problem is **stable** if all time-dependent solutions starting out sufficiently close to the equilibrium stay in any given neighbourhood of it, for all time $t > 0$. Moreover, if all such solutions approach the equilibrium as $t \to \infty$, we say that the equilibrium is **asymptotically stable**. An equilibrium is **unstable** if we can find solutions starting arbitrarily close to it which leave some given small neighbourhood as $t \to \infty$. Here, expressions such as *close to, approaching* and *neighbourhood* must be defined in terms of some suitable metric, or norm, on the function space containing solutions.

Suppose $u_0(\mathbf{x})$ is an equilibrium for the problem,

$$u_t = \Delta u + f(u, \nabla u, \mathbf{x}), \quad \mathbf{x} \in \Omega,\ t > 0s, \tag{1.4.10}$$

with appropriate boundary conditions if $\Omega \neq \mathbf{R}^n$. Set $u(\mathbf{x},t) = u_0(\mathbf{x}) + v(\mathbf{x},t)$, where v is initially, at least, a small perturbation (i.e. $v(\mathbf{x},0)$ is arbitrarily small). Then, substituting into (1.4.10), and retaining only linear terms in v, we have

$$v_t = \Delta v + \sum_{i=1}^{n} f_{u_{x_i}}(u_0, \nabla u_0, \mathbf{x}) v_{x_i} + f_u(u_0, \nabla u_0, \mathbf{x}) v. \tag{1.4.11}$$

If boundary conditions are imposed on u, then the appropriate conditions must also be applied to v.

The equation (1.4.11), together with any boundary conditions, constitutes a linear problem, and may be written formally as

$$v_t + Av = 0, \tag{1.4.12}$$

where A is a linear differential operator defined on a suitable class of functions, which satisfy the boundary conditions, so that $-Av$ agrees with the right-hand side of (1.4.11).

The domain of A, $D(A)$, may be extended to be a dense subset of some Banach space, X, such as $L_2(\Omega)$ (see the example below or Box D).

Now the terms used in defining stability, etc., can be made precise in the X-norm.

The equilibrium u_0 is **asymptotically stable** if $\|v\|_X \to 0$ as $t \to \infty$. We can guarantee this when the **spectrum** of A can be contained strictly in the right-hand side of the complex plane – that is there is $\beta > 0$ such that $Re\ (\lambda) \geq \beta$ whenever $\lambda \in \sigma(A)$, the spectrum of A. In this case, it can be shown that $\|v\|_X$ decays like $e^{-\beta t}$. We refer the reader ahead to Box D for any definitions concerning spectral theory that may be required at this stage.

For simple initial-boundary-value problems on compact domains, the spectrum of our operator A consists of a sequence of eigenvalues, $\{\lambda_m\}$, each one being associated with an eigenfunction ϕ_m, such that

$$A\phi_m = \lambda\phi_m.$$

Suppose $Re(\lambda_m) \leq 0$ for some m. Then if

$$v(\mathbf{x}, 0) = (\text{constant})\phi_m(\mathbf{x}),$$

the solution of (1.4.12) is simply

$$v(\mathbf{x}, t) = (\text{constant})e^{-\lambda_m t}\phi_m(\mathbf{x}),$$

which does not approach zero as $t \to \infty$.

The converse is also true, since it is known that the eigenfunctions span the function space containing the solutions, v. Hence any solution may be expanded as a (Fourier) series:

$$v(\mathbf{x}, t) = \sum_{m=1}^{\infty} a_m(t)\phi_m(\mathbf{x}). \tag{1.4.13}$$

Now projecting (1.4.12) on to the linear subspaces spanned by each of the $\phi_m(\mathbf{x})$, we see that the coefficients $a_m(t)$ in (1.4.13) satisfy

$$\frac{da_m}{dt} + \lambda_m a_m = 0.$$

Thus if $v \to 0$ as $t \to \infty$, we must have $\lambda_m < 0$, for each m.

Similar considerations apply when the domain Ω is unbounded. Generally, the **spectrum** of A not only contains discrete **eigenvalues** but also a continuum, known as the **essential spectrum** (see Box D). The location of the essential spectrum depends very delicately upon the choice of function space containing possible initial values for v. (In fact, if this function space were restricted greatly, some eigenvalues might also disappear!) In Box D, we give some general examples of operators and their spectral properties and an explicit method to calculate the essential spectrum for systems like (1.4.11), when $\Omega = \mathbf{R}$.

Example

Consider the example given by (1.4.10). Here we saw nonpositive equilibria bifurcate from $u \equiv 0$ as L increased through π. Linearizing (1.4.10) about $u \equiv 0$, we get

$$v_t = v_{xx} + v, \quad x \in (0,1), \quad v(0) = v(L) = 0. \tag{1.4.14}$$

Our operator A is defined by

$$Av = -v_{xx} - v$$

for C^2 functions v satisfying $v(0) = v(L) = 0$. Using the L_2 norm, A may be extended so that its domain $D(A)$ is dense in $L_2(0, L)$, and as large as possible.

Since the eigenfunctions of the operator $-\frac{d^2}{dx^2}$ in $L_2(0, L)$, with homogeneous Dirichlet boundary conditions, are of the form $\sin \frac{n\pi x}{L}$, it is easy to see that the eigenvalues of A are given by

$$\lambda_n = -1 + \frac{n^2\pi^2}{L^2}, \quad n = 1, 2, \ldots,$$

each with eigenfunction $\phi_n = \sin \frac{n\pi x}{L}$. Notice that, in this case, (1.4.13) is simply the Fourier sine series for v and the stability analysis simply involves the usual *separation of variables.* In practice, for such examples, we do not need to introduce the operator A, but rather, we may proceed directly from (1.4.14) to a calculation of the eigenfunctions for the operator on the right-hand side. Now notice that $0 < \lambda_1 < \lambda_2 \ldots$, provided $L^2 < \pi^2$; $\lambda_1 = 0$ when $L = \pi$ and $\lambda_1 < 0$ for $L > \pi$. Thus the stability of $u \equiv 0$ changes from asymptotically stable to unstable as L increases through π. It is no coincidence that a bifurcation of equilibria takes place when $L = \pi$. It is always the case that bifurcation points are associated with a change of stability for one or more of the equilibria involved. As a further example, consider the following problem (an example of which we solved numerically in section 1.3):

$$u_t = u_{xx} + f(u), \quad x \in (0, 10), \quad u_x(0) = u_x(10) = 0,$$

where f is some smooth function with a root at u_0, say. Clearly $u \equiv u_0$ is an equilibrium.

Linearizing about $u \equiv u_0$ we have

$$v_t = v_{xx} + f_u(u_0)v, \quad x \in (0, 10), \quad v_x(0) = v_x(10) = 0.$$

Now the eigenvalue problem

$$-\phi_{xx} - f_u(u_0)\phi = \lambda\phi, \quad \phi_x(0) = \phi_x(10) = 0,$$

has solutions

$$\phi_n = \cos\frac{n\pi x}{10}, \quad \lambda_n = -f_u(u_0) + \frac{n^2\pi^2}{100}, \quad n = 1, 2, \ldots.$$

Thus, if $f_u(u_0)$ is negative, the equilibrium is asymptotically stable, while if $f_u(u_0)$ is greater than $\pi^2/100$, it is unstable (certainly to perturbations in the first mode, $\phi_0 \equiv 1$).

In Chapter 2, we shall show how systems of equations can display subtle stability and bifurcation behaviour (not just involving the simplest eigenmodes).

We close this section with an example where we do not evaluate the equlibria or solve the eigenvalue problem explicitly.

Example

Consider

$$u_t = u_{xx} - uu_x, \; x \in (0, L), \; t > 0, \quad u(0) = 0, \; u(L) = 1.$$

An equilibrium, u^*, satisfies

$$u^*_x = \frac{u^{*2} + \alpha}{2},$$

for some real constant α. Thus if $\alpha \geq 0$, applying the boundary condition at 0, we have

$$u^* = \sqrt{\alpha}\tan(\sqrt{\alpha}.x/2).$$

Hence if $\alpha \geq 0$ satisfies $1 = \sqrt{\alpha}.\tan(\sqrt{\alpha}L/2)$, we have an equilibrium solution (e.g. if $L = \pi/2$, then choose $\alpha = 1$). Suppose this is so; then linearizing about u^*, we have

$$v_t = v_{xx} - (vu^*)_x, \; x \in (0, L), \; t > 0, \quad v(0) = v(L) = 0.$$

Eigenfunctions of the operator on the right-hand side satisfy,

$$\lambda\phi = -\phi_{xx} + (\phi u^*)_x, \quad x \in (0, L), \quad \phi(0) = \phi(L) = 0.$$

Multiplying this by ϕ and integrating over $(0, L)$;

$$\begin{aligned}
\lambda \int_0^L \phi^2 \, dx &= -\int_0^L \phi_{xx}\phi \, dx + \int_0^L \phi(\phi u^*)_x \, dx \\
&= \int_0^L \phi_x^2 \, dx + \int_0^L \phi(\phi u^*)_x \, dx \\
&= \int_0^L \phi_x^2 \, dx - \int_0^L \left(\frac{\phi^2}{2}\right)_x u^* \, dx \\
&= \int_0^L \phi_x^2 \, dx + \int_0^L \frac{\phi^2}{2} u_x^* \, dx.
\end{aligned}$$

But $u_x^* > 0$ on $(0, L)$, so the right-hand side is strictly positive for all $\phi \not\equiv 0$. Hence all eigenvalues must be positive, and u^* is asymptotically stable.

Box D: The spectra of some differential operators

Let X be a Banach space (a complete, normed vector space) and $A : D(A) \to X$ be a linear operator with domain $D(A) \subseteq X$. For any complex number λ we form the operator

$$A_\lambda = A - \lambda I,$$

where I is the identity operator on X.

If A_λ has an inverse, we denote it by

$$R_\lambda(A) = A_\lambda^{-1}.$$

It is called the resolvent of A.

Now λ is a **regular** point for A if:

(i) $R_\lambda(A)$ exists,
(ii) $R_\lambda(A)$ is bounded,
(iii) $R_\lambda(A)$ is defined on a dense subset of X.

The **point spectrum** or **eigenvalues** are values of λ for which $R_\lambda(A)$ does not exist.

The **continuous spectrum** is the set of values of λ for which $R_\lambda(A)$ exists but is unbounded.

The **residual spectrum** is the set of values of λ for which $R_\lambda(A)$ exists, is bounded, but is not defined on a dense subset of X.

We let $\sigma(A)$ denote the **spectrum** of A (i.e. all nonregular points), and say that the **essential spectrum** of A is any point in $\sigma(A)$ which is not an isolated eigenvalue of finite multiplicity. (Such eigenvalues are surrounded by a neighbourhood of regular points and are such that the dimension of the null space of A_λ is finite.)

Example

Let $X = L_2(\mathbf{R})$. For twice continuously differentiable functions $u(x)$, with u_{xx} in X, we define

$$Au = -u_{xx} + 2au_x + bu, \tag{1}$$

where a and b are constants.

We may extend the definition of A so that it becomes a **closed** operator with domain, $D(A)$, dense in X. (That is, if $\{x_n\}$ is a sequence in X converging to x, such that the sequence $\{Ax_n\}$ also converges in X, to y, say, then we define $Ax = y$ and include x in $D(A)$.) The spectral theory of closed linear operators is well developed (see [42] for example). However, we shall locate the spectrum of A directly, as follows.

Firstly, suppose A_λ is not invertible for some λ; then there exists an eigenfunction $w \in X$ such that $A_\lambda w = 0$. This last equation is linear with constant coefficients. Hence, any solution cannot decay at both $\pm\infty$, and so cannot be in X. Thus A has no eigenvalues and A_λ always possesses some kind of inverse, R_λ. We may represent this by means of a Green function, G. Let $h \in D(R_\lambda) \subseteq X$. Then we may write

$$u = R_\lambda h$$

as

$$u = \int_{\mathbf{R}} G(x - s)h(s)\, ds,$$

where G is of the form

$$G(y) = \begin{cases} \alpha \exp(\mu_1 y) & y \leq 0 \\ \alpha \exp(\mu_2 y) & y \geq 0, \end{cases}$$

the μ_i are roots of the characteristic polynomial,

$$P(\mu) = \mu^2 - 2a\mu + \lambda - b,$$

and α is chosen so that $-1 = \alpha(\mu_2 - \mu_1)$.

Now Young's inequality [22] asserts that

$$\|f * g\|_{L_r} \leq \|f\|_{L_p} . \|g\|_{L_q},$$

where

$$\frac{1}{r} = \frac{1}{p} + \frac{1}{q} - 1$$

so long as $f \in L_p$ and $g \in L_q$. So we have

$$\|u\|_{L_2} \leq \|G\|_{L_1}.\|h\|_{L_2},$$

whenever $\|G\|_{L_1} < \infty$.

If P is such that μ_1 has positive real part, and μ_2 has negative real part, then $G \in L_1$, trivially. Thus, in this case, R_λ is clearly bounded with dense domain equal to X.

The roots of P depend continuously upon λ. For λ real $< b$, P possesses two real roots, one positive, one negative. As λ varies over the complex plane, roots may only cross the imaginary axis, at ik say, when

$$\lambda = b + k^2 + 2aki.$$

Hence, for λ to the left of this parabola: that is, where

$$Re(\lambda) < b + Im(\lambda)^2/4a^2,$$

P has a root on either side of the imaginary axis. Thus

$$\sigma(A) \subseteq \{\lambda : Re(\lambda) \geq b + \frac{Im(\lambda)^2}{4a^2}\},$$

and, in fact, includes the parabola itself. To see this, let $\psi(x)$ be some smooth function with $\psi = 1$ for $|x| < 1$, and $\psi = 0$ for $|x| > 2$. For integers, m, set $u_m = \psi(\frac{x-3m}{m})e^{ikx}$, where $\lambda = b + k^2 + 2aki$. Then, by direct calculation, one can show

$$\frac{\|A_\lambda u_m\|_{L_2}}{\|u_m\|_{L_2}} = O(m^{-1})$$

as $m \to \infty$. Hence R_λ is unbounded here, so $\lambda \in \sigma(A)$.

For general second-order systems, with nonconstant coefficients, the following result, from [35], is useful.

Theorem. Suppose M and N are bounded real $(m \times m)$ matrix functions of $x \in \mathbf{R}$, such that $M(x), N(x) \to M_\pm, N_\pm$ as $x \to \pm\infty$. Suppose that D is a constant, symmetric, positive definite matrix. Then we define

$$Au = -Du_{xx} + Mu_x + Nu$$

as a closed, densely defined operator on $L_2(\mathbf{R}, \mathbf{R}^m)$.

Let

$$S_{\pm} = \{\lambda : \ \det(k^2 D + ikM_{\pm} + N_{\pm} - \lambda I) = 0 \ ; \ k \in \mathbf{R}\}.$$

Then $S_{\pm}$ consists of a finite number of curves, each parameterised by k, symmetric about the real axis, and asymptotic to a parabola.

Let Q denote the union of the regions inside and on the curves S_+, S_-. Then the essential spectrum of A lies in Q and, in particular, includes S_+ and S_-.

Note that linear differential operators on $L_2(\mathbf{R})$ with nonconstant coefficients may possess eigenvalues besides the essential spectrum.

So far, we have considered our operators A to be defined in the Banach space $L_2(\mathbf{R})$. There is a similar theory when differential operators are defined in L_p spaces or C^0. In fact, the above theorem concerning the essential spectra still holds good for these spaces.

To see that the choice of underlying space is important, we consider the weighted L_2 space defined as follows. Let

$$\|u\|_Y = \{\int_{\mathbf{R}} |u(x)|^2 \exp 2\gamma x \ dx\}^{1/2},$$

and define Y to be the space of functions where $\|u\|_Y < \infty$.

Consider the operator A, defined by (1), as a closed, densely defined operator in Y. We shall denote it A^* to stress the difference between its domain and that of A, which was defined in X. Clearly, for $u \in Y$, we have $v = u \exp \gamma x$ in $L_2(\mathbf{R})$. Thus

$$A^*_\lambda u = h$$

in Y is equivalent to

$$\exp \gamma x . A^*_\lambda \left(\frac{v}{\exp \gamma x} \right) = g \ (= \cosh \gamma x . h)$$

in X. Define

$$B = \exp \gamma x . A^* \left(\frac{\cdot}{\exp \gamma x} \right),$$

which is the operator on X induced by A^*. The spectrum of A^* corresponds exactly to the spectrum of B. Formally

$$Bv = -v_{xx} + (2a + 2\gamma) v_x + (-\gamma^2 - 2a\gamma + b) v.$$

Now the essential spectrum of B, and hence of A^*, is contained in the half-plane

$$Re(\lambda) \geq b - \gamma^2 - 2a\gamma.$$

Thus when γ and a are of opposite signs and

$$|\gamma| < 2|a|,$$

the spectrum of A^* lies further to the right in the complex plane than that of A.

The ability of weighted spaces to push the essential spectrum of a differential form around is important since we may naturally wish to consider more restricted perturbations in a linearized stability problem. By forcing the perturbations to lie in Y, say, we may be able to prove a stability result unavailable in X. This is particularly useful when we want to move the essential spectrum by only an arbitrarily small amount (see the examples in sections 3.2 and 5.4).

It is important to realize that where an operator on X has isolated eigenvalues, if we consider its restriction to Y, then the eigenvalue will vanish as soon as a weighted space demands decay at infinity faster than that of the associated eigenfunction. This may well mean that we have disallowed intuitively reasonable functions from the new domain of our operator.

So far, we have considered functions defined on $\mathbf{R}$. Similar considerations apply when Ω is a more general noncompact subset of $\mathbf{R}^n$. We pass over this here for the sake of brevity.

When our domain Ω is compact, and simple linear boundary conditions are applied at $\partial\Omega$, there is a well developed spectral theory. Specifically, the spectrum consists solely of eigenvalues, which, for closed operators, are all of finite multiplicity. It may not be possible to construct the eigenvalues explicitly though. Readers are referred to [51] or [22], and the examples in the main text.

1.5 A travelling wave

One of the most important properties of nonlinear parabolic systems is their ability to support travelling wave solutions. Unlike the linear wave equation, for example, which is hyperbolic and propagates any wave profile with a specific speed, reaction-diffusion equations may only allow certain wave profiles to propagate, each one with its own characteristic velocity.

The construction and study of wave solutions for nonlinear parabolic systems is an area of great interest, not only in the applications of the waves themselves, but also in their use in gaining a better understanding of phenomena in large domains, where, locally at least, the patterns may resemble travelling waves.

Until recent years, almost all of the attention was focused on problems concerning plane waves, that is waves moving through $\mathbf{R}^n$ in a fixed direction, having translational symmetry in any direction perpendicular to that of the motion. In this section and Chapter 3, we shall be concerned with such waves, and shall try to indicate techniques which enable us to analyse them qualitatively.

However, such waves are only the beginning of the story. For some time, chemists and others have observed many strange wave-like structures in systems that could be modelled via reaction-diffusion equations. Spiral waves and scroll waves have become well documented within certain chemical reactions [77]. The problem was how to approach such objects mathematically. Over the last two years or so, an answer (at least in part) has been developed. This is the content of Chapter 4. There, we shall show how one can *wrap up* plane waves on to evolving surfaces in order to obtain many exotic wave-type phenomena. This exciting theory takes the construction of plane waves (or one-dimensional waves) as its starting point, and assumes that such problems have already been solved. Thus, our current task is even more important and I would suggest that plane wave problems are the most fundamental problems in reaction-diffusion.

Consider the equation

$$u_t = \Delta u + f(u). \tag{1.5.1}$$

Here f is of the form depicted in Figure 1.7, and has roots at $u = 0$, a, and 1.

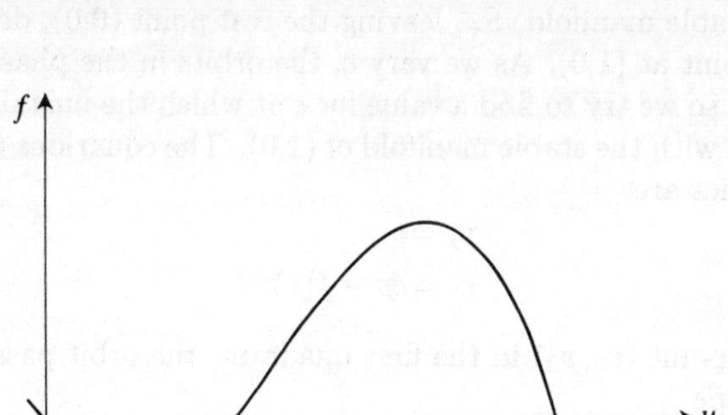

Figure 1.7: f in (1.5.1).

We seek a solution in the form of a travelling wave. That is, in the form

$$u(\mathbf{x}, t) = v(z), \tag{1.5.2}$$

where

$$z = \mathbf{n}.\mathbf{x} + ct.$$

Here c is some constant, the wave speed, to be determined, while $\mathbf{n}$ is a unit vector. A solution of the form (1.5.2) is a wave with profile given by v, moving with speed c in the direction $-\mathbf{n}$.

Substituting (1.5.2) into (1.5.1), we arrive at the ordinary differential equation:

$$v_{zz} - cv_z + f(v) = 0. \tag{1.5.3}$$

Now (1.5.1) possesses two stable equilibria: $u = 0$ and $u = 1$. We shall seek a wave v which connects these two states, so that one is replaced by the other as the wave propagates over $\mathbf{R}^n$. Thus, we solve (1.5.3) subject to

$$v \to 0, \text{as } z \to -\infty, \quad v \to 1, \text{as } z \to \infty.$$

First suppose $c = 0$. Then solutions of (1.5.3) satisfy

$$\frac{v_z^2}{2} + \int_0^v f(w)\, dw = \text{constant}.$$

We may sketch the solutions as orbits in the (v, v_z)-phase plane, see Figure 1.8. Here, we have assumed that $\int_0^1 f(w)\, dw > 0$ (if this is not so, we may transform $u \to 1 - u$, and proceed as before).

The unstable manifold, S_+, leaving the rest point (0,0), does not reach the saddle point at (1,0). As we vary c, the orbits in the phase plane vary continuously, so we try to find a value for c at which the unstable manifold, S_+, coincides with the stable manifold of (1,0). The equations for the phase plane dynamics are

$$\begin{aligned} v_z &= p \\ p_z &= cp - f(v). \end{aligned}$$

Now, at any point (v_0, p_0) in the first quadrant, the orbit passing through it has the slope

$$c - \frac{f(v_0)}{p_0}$$

which is monotonically increasing with c. Thus, one can see that the unstable manifold, S_+, moves up above its previous position in the manner depicted in Figure 1.9, as c increases.

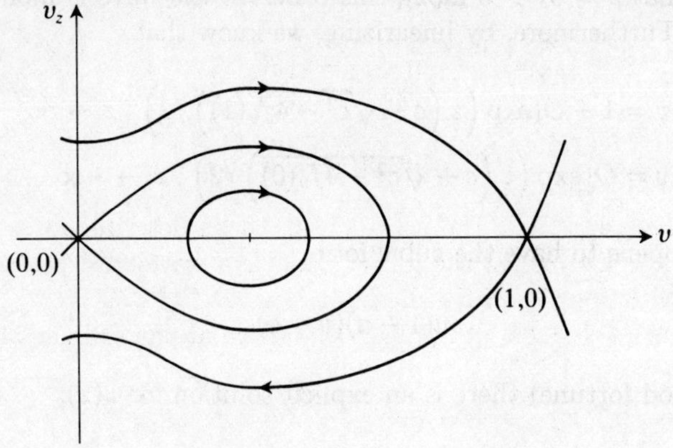

Figure 1.8: Phase plane for (1.5.3) when $c = 0$.

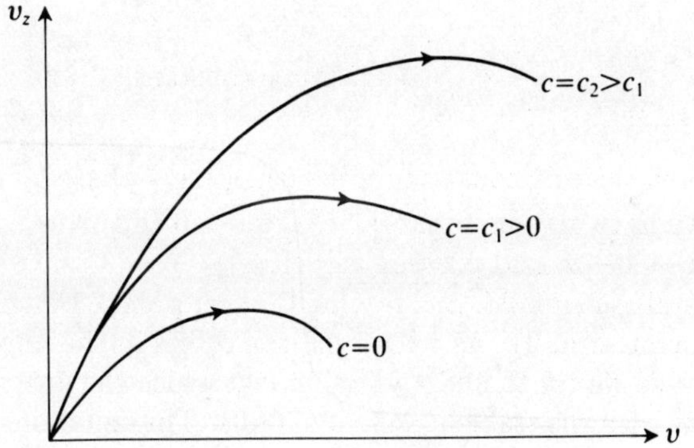

Figure 1.9: The location of S_+ as c increases from 0.

Clearly, there will be a unique value of $c > 0$ when S_+ reaches (1,0). For greater values of c, S_+ never reaches the line $v_z = p = 0$. For lesser values of c, S_+ never reaches $v = 1$. Thus, we have a unique solution.

Also note that $p = v_z > 0$ along this orbit, so the wave is monotonically increasing. Furthermore, by linearizing, we know that

$$v = 1 + O(\exp\left(z\left(c - \sqrt{c^2 - 4f_v(1)}\right)/2\right) \quad z \to \infty$$
$$v = O(\exp\left(z\left(c + \sqrt{c^2 - 4f_v(0)}\right)/2\right) \quad z \to -\infty$$

If f happens to have the cubic form,

$$u(1-u)(u-a),$$

then (by good fortune) there is an explicit solution for $v(z)$:

$$v(z) = \left(1 + \exp\left(-z/\sqrt{2}\right)\right)^{-1},$$

where $c = \sqrt{2}(0.5 - a)$.

Our solution is depicted in Figure 1.10. The wave shifts the state variable from the zero state to the unitary state as it propagates.

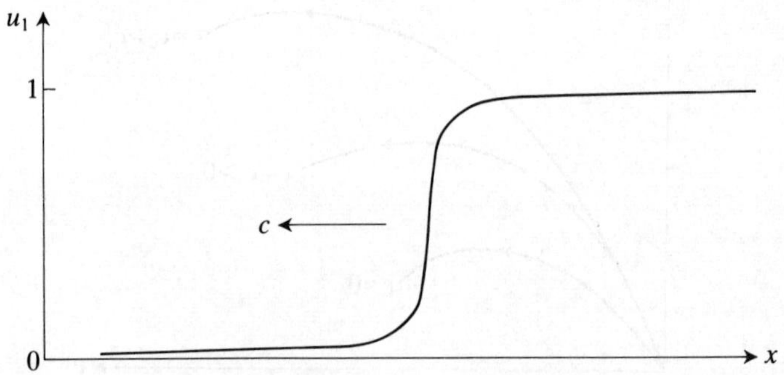

Figure 1.10: The travelling wave-front solution for (1.5.1).

Finally, before we leave this example, we note that, from (1.5.3), we can obtain

$$c\int_{\mathbf{R}} v_z^2\, dz = \int_0^1 f(v)\, dv.$$

Thus, in the general case, the sign of c is determined by $\int_0^1 f(u)\, du$. This is useful since, if f depends on a number of parameters and the sign of

c is known (better still if c is known to be zero!), we have a generic condition which restricts the parameters employed describing the nonlinearity f. In the above example, we were able to employ a simple qualitative phase plane argument in order to deduce the existence and uniqueness of the wave. In Chapter 3, we shall consider extensions of these and other ideas which may be employed in the construction of plane wave solutions.

Not all systems have travelling wave properties as clear cut as (1.5.1). We briefly consider the following well-known case.

Example

Consider the following equation, usually called Fisher's equation:

$$u_t = u_{xx} + u(1-u).$$

Looking for travelling wave solutions $u = v(z)$, $z = x + ct$, we have the phase plane system

$$v_z = p$$
$$p_z = cp - u(1-u).$$

Linearizing at (0,0), we see that the origin is unstable for all $c > 0$. Orbits spiral out from the rest point for $2 > c$, and enter the first quadrant (without spiralling) for $c \geq 2$.

Now, for $c \geq 2$, it is easy to check that no orbits can enter the triangle, T, given by

$$0 \leq v \leq 1, \quad 0 \leq p \leq \frac{cv}{2},$$

from outside (check the slopes of the orbits at points on the boundaries). Thus, all orbits inside T must emanate from the origin. However, (1,0) has a stable manifold inside T (linearize about the rest point). Hence, it must connect (0,0) to (1,0) inside T.

We have shown that for all wave speeds $c \geq 2$, there exists a monotone wave-front solution of our equation, connecting the states $u = 0$ and $u = 1$. No such solution may exist for $c < 2$.

Box E: Group invariant solutions

Travelling wave solutions of autonomous partial differential equations are simple examples of **group invariant** solutions (otherwise known as **similarity** solutions). The idea is to construct solutions of a partial differential equation by reducing the problem to a system involving fewer independent variables.

For example, when we seek plane travelling wave solutions of an equation of the form

$$G(u_t, u, u_x, u_{xx}) = 0,$$

we simply substitute $u = v(z)$, where $z = x+ct$. We then solve the resulting ordinary differential equation

$$G(cv_z, v, v_z, v_{zz}) = 0$$

for v, exploiting the wave speed c in order to seek bounded solutions. The resulting solution $u = v(z)$ may be pictured as a surface in $\mathbf{R}^3$ (i.e. (x, t, u)-space). Clearly, this surface is unchanged by the transformation

$$x \to x - \varepsilon c, \ \ t \to t + \varepsilon, \ \ u \to u, \tag{1}$$

for any real constant ε. We say that the solution is invariant under the **group** of such transformations. For each ε fixed, (1) defines the corresponding group element. (Convince yourself that this is indeed a group!).

Travelling wave solutions can always be sought provided the original equation contains no explicit dependence upon x and t. In this case, the transformation (1) maps **any** solution on to another solution.

It is natural to ask whether this kind of method works for other groups of transformations. The answer is yes. We outline the method below as applied to equations with one dependent variable and two independent variables.

Consider a partial differential equation of the form

$$G(x, t, u, u_t, u_x, \ldots) = 0. \tag{2}$$

Here, a solution $u(x, t)$ should be visualized as a surface in $\mathbf{R}^3$, (x, t, u)-space. Now suppose that we have a one-parameter group of transformations for $\mathbf{R}^3$ as follows:

$$\begin{aligned} x &\to X(x, t, u, \varepsilon) \\ t &\to T(x, t, u, \varepsilon) \\ u &\to U(x, t, u, \varepsilon) \end{aligned} \tag{3}$$

defined for all $\varepsilon \in \mathbf{R}$, such that

$$X \equiv x, \ T \equiv t, \ U \equiv u, \text{ when } \varepsilon = 0.$$

The group of transformations (3) is called a **symmetry group** for (2), if each element in (3) maps any solution of (2) on to another solution of (2).

For example, the group of transformations,

$$\begin{aligned} x &\to \varepsilon x \\ t &\to \varepsilon^2 t \\ u &\to \varepsilon^{\alpha} u \end{aligned} \tag{4}$$

(for any fixed constant α), is a symmetry group for the diffusion equation

$$u_t = u_{xx}. \tag{5}$$

(To check, set $U = \varepsilon^\alpha u,\ X = \varepsilon x,\ T = \varepsilon^2 t$ and show that $U_T = U_{XX}$ whenever u satisfies (5).)

Returning to the more general problem, we ask whether the problem (2) has any solutions which are **invariant** under the group of transformations (3). Any such solution is a surface in $\mathbf{R}^3$ which remains unaltered by the transformations (3). More exactly, we would require the curve

$$\{(X(x,t,u,\varepsilon), T(x,t,u,\varepsilon), U(x,t,u,\varepsilon)) :\ \varepsilon \in \mathbf{R}\}$$

to lie in the surface whenever (x,t,u) does. For fixed (x,t,u), we refer to the set

$$\{(X(x,t,u,\varepsilon), T(x,t,u,\varepsilon), U(x,t,u,\varepsilon)) :\ \varepsilon \in \mathbf{R}\}$$

as the **orbit** through (x,t,u).

Now we suppose that our transformation (3) has two **algebraic invariants**, i.e. two functionals of the form

$$y = y(x,t,u); \quad w = w(x,t,u), \tag{6}$$

which are left unchanged along orbits (i.e. are independent of ε when (3) is applied).

In our example (4), we can choose

$$y = \frac{x^2}{t}, \quad w = ut^{-\alpha/2}. \tag{7}$$

Any surface, S, which is invariant under the group of transformations (3), can be specified by defining a relationship between the invariants w and y. It is clear that such a relation defines some surface in $\mathbf{R}^3$, via (6), which contains the whole orbit of every point in it (and hence, is invariant). Conversely, if we have an invariant surface, S, we may obtain the graph of w versus y by evaluating them upon some curve, traversing all the orbits embedded in S. We shall regard the defining relationship to be of the form

$$w = w(y)$$

so that y may be thought of as a new independent variable and w as a new dependent variable.

Now comes the neat part. Let us assume that (6) is invertible for u and x in terms of y, w, and t, so that we have

$$\begin{aligned} u &= a(y,w,t) \\ x &= b(y,w,t). \end{aligned} \tag{8}$$

Next, we calculate the derivatives of u by applying the chain rule to the expressions (8), and using (6). (We write all the derivatives of u as functions of $y,\ w,\ t$ and the ordinary derivatives $w_y,\ w_{yy},\ldots$.)

In our example, from (7),

$$u = t^{\frac{\alpha}{2}} w, \quad x = \sqrt{yt},$$

so that,

$$\begin{aligned} u_t =& \frac{\alpha}{2} t^{\frac{\alpha}{2}-1} w + t^{\frac{\alpha}{2}} w_y y_t \\ =& \frac{\alpha}{2} t^{\frac{\alpha}{2}-1} w + t^{\frac{\alpha}{2}} w_y \frac{-x^2}{t^2} \\ =& t^{\frac{\alpha}{2}-1} \left(\frac{\alpha}{2} w - w_y y \right). \end{aligned} \tag{9}$$

Similarly,

$$u_x = t^{\frac{\alpha}{2}} w_y y_x$$

and

$$u_{xx} = t^{\frac{\alpha}{2}-1}(2w + 4y w_{yy}). \tag{10}$$

Now it turns out that, if (3) is a symmetry group for (2), then, when we substitute the expressions in terms of $t,\ y,\ w,\ w_y,\ w_{yy}$, etc., for $x,\ u,\ u_t,\ u_x$, etc. (in (2)), the t-dependence of the resulting expression can be cancelled out. Thus we obtain an ordinary differential equation for $w(y)$. If this can be solved satisfactorily, then (8) and (6) provide the solution in (x, t, u)-space, as required.

Returning to our example, if u satisfies (5), then (9) and (10) imply

$$\frac{\alpha}{2} w - y w_y = 2 w_y + 4 y w_{yy}. \tag{11}$$

Here we have cancelled $t^{\frac{\alpha}{2}-1}$ throughout. Recall that α was arbitrary in (4), so let us now choose it to be -1. Then (11) becomes

$$4y(w_y + \frac{w}{4})_y + 2(w_y + \frac{w}{4}) = 0.$$

Clearly $w = \exp(-y/4)$ is a solution, so using (7) we obtain a solution of (5):

$$u = t^{-\frac{1}{2}} \exp\left(\frac{-x^2}{4t}\right),$$

the fundamental solution of the diffusion equation.

The connection between group invariance (or symmetries) of differential equations and their possible simplification (or integrability) is an elegant one. In Hamiltonian dynamics, for example, any symmetry in the system provides first integrals, which reduce the order of the problem, and bring us closer to solving it.

As a topic in nonlinear parabolic equations, the more esoteric group invariant solutions tend to be of less importance. When invariant solutions exist, and can perhaps be found explicitly, they are useful touchstones, providing some insight into the kind of behaviour to be expected when general initial-value problems are considered.

Still, there is nothing lost in playing with these ideas for a while. A few moment's experimentation with one's favourite equation may yield interesting results.

We close with one more example.

Example

Consider

$$u_t = u_{xx} + uu_x,$$

which is invariant under the group,

$$x \to x - \varepsilon t, \quad t \to t, \quad u \to u + \varepsilon.$$

We choose invariants $y = t$ and $w = u + \frac{x}{t}$. Then $w(y)$ satisfies,

$$w_y = -\frac{w}{y}.$$

Thus $w = (\text{constant})/y$, and so

$$u = \frac{(C - x)}{t}$$

is a solution of the original equation for any constant C.

1.6 Local existence theory

Although these notes are primarily concerned with techniques for applied mathematicians, we pause here to consider the theoretical aspect of well-posedness (i.e. whether we have specified a sensible problem which possesses a solution).

Those who are interested in techniques for analysing more practical problems in reaction-diffusion are welcome to skip over this section. However, for those who arrive at reaction-diffusion systems by modelling processes in the applied sciences, I would suggest that a few moments' thought along the lines of the material in this section (before attempting any applied or numerical analysis) may save both time and frustration later.

For most reaction-diffusion systems, local existence theory is not too much of a problem. Hopefully, a consideration of the examples below would be sufficient to illustrate the kind of pitfalls involved.

By **local existence** of solutions, we mean existence of solutions of initial-boundary-value problems for a small time interval. We reserve the term **global existence** for problems where we can show that solutions exist for all time after the imposition of the initial conditions.

From the start, I must concede that we do not have the space to do justice to this elegant theory, so the interested reader is referred to Dan Henry's lecture notes [35] where the general theory is developed (as well as much more).

We shall skim the surface of existence theory, stating the main theorem and considering some examples. The basic ideas behind this section are from functional analysis: we shall develop these briefly and state an embedding theorem which allows us to proceed at a pace through our examples.

The fundamental point is that when reaction-diffusion systems are viewed as evolution equations in some function space (i.e. for each time, t fixed, $u(t)$ is in a given function space), the equation may be thought of as an ordinary differential equation. Then, the theory of existence and uniqueness for initial-value problems apes that for ordinary differential equations (Picard's theorem etc.). The difference is that our basic space is a function space rather than a finite-dimensional Euclidean vector space. Hence, we must make sure that mappings take us into domains of certain linear operators. These considerations never arise, of course, in finite-dimensional spaces.

As indicated earlier in Box D, we define our differential operators in spaces chosen in such a way as to incorporate any boundary conditions. This naturally affects the spectral properties, but the theory is quite robust and applies equally well to systems on bounded or unbounded domains.

Consider

$$u_t = \Delta u + f(\mathbf{x}, u, \nabla u), \quad \mathbf{x} \in \Omega \subseteq \mathbf{R}^n, \ t > 0. \tag{1.6.1}$$

If $\Omega \neq \mathbf{R}^n$, then we impose some boundary conditions of the form

$$\alpha(\mathbf{x})u + \beta(\mathbf{x})\nabla u.\mathbf{n} = 0, \quad \mathbf{x} \in \partial\Omega, \ t > 0, \tag{1.6.2}$$

where α and β are given, and $\mathbf{n}$ is the outer normal to Ω.

To complete the problem we specify some initial conditions:

$$u(\mathbf{x}, 0) = u_0(\mathbf{x}), \quad \mathbf{x} \in \Omega. \tag{1.6.3}$$

We shall develop an existence theory for solutions of (1.6.1)-(1.6.3) in $L_2(\Omega)$). This necessarily implies some decay if $|\mathbf{x}| \to \infty$ in Ω. The theory may be presented in less restrictive spaces, such as $C(\Omega)$, but we pass over this here and refer to [35].

Now for smooth functions in $L_2(\Omega)$ satisfying the boundary conditions (1.6.2), we define

$$Au = -\Delta u$$

and extend A to be a closed, densely defined operator in $L_2(\Omega)$ (see Box D, for example).

We can rewrite (1.6.1)-(1.6.3) as:

$$\begin{aligned} u_t + Au &= f, \quad t > 0, \\ u &= u_0, \quad t = 0. \end{aligned} \tag{1.6.4}$$

Notice that any boundary conditions have been absorbed into the definition of the domain of A, $D(A)$.

The theory we present holds for a general class of operators A, not simply the case illustrated above. We make this generalization here and shall give an existence theory for (1.6.4), (which includes (1.6.1)-(1.6.3) as an example). Specifically we assume that: A is a closed, densely defined operator in a Banach space X, and there are constants $a \in \mathbf{R}$, $\phi \in (0, \pi/2)$, $M \geq 0$ such that the sector

$$\{\lambda : \ \phi \leq |\arg(\lambda - a)| \leq \pi, \ \lambda \neq a\}$$

in the complex plane contains no part of the spectrum of A, and that the estimate

$$\|(\lambda I - A)^{-1}\| \leq \frac{M}{|\lambda - a|}$$

holds in this sector.

Such operators are called **sectorial operators.**

All of the operators that we have met have been sectorial (one may check this directly for most of them using the ideas developed in Box D in section 1.4). In all cases, we have used $X = L_2(\Omega)$.

Before stating our existence theorem, we introduce some useful function spaces. For a sectorial operator A, we set $A_1 = A + a$ so that its spectrum, $\sigma(A_1)$, lies in the right half of the complex plane. Now for $\alpha \geq 0$, we define the operator

$$A_1^{-\alpha} = \frac{1}{\Gamma(\alpha)} \int_0^\infty t^{\alpha-1} \left(\frac{1}{2\pi i} \int_C (\lambda + A_1)^{-1} \exp(\lambda t) \, d\lambda \right) dt,$$

where C is a contour to the right of $\sigma(-A_1)$ with $\arg \lambda \to \pm\theta$ as $|\lambda| \to \infty$, for some $\theta \in (\pi/2, \pi)$. This is admittedly a little technical, but we only need to observe some of the properties of the $A_1^{-\alpha}$.

It can be shown that, for $\alpha > 0$, $A_1^{-\alpha}$ is a bounded one-to-one operator on X. So we set

$$A_1^{\alpha} = \text{inverse of } A_1^{-\alpha}, \quad \alpha > 0,$$
$$A_1^0 = \text{the identity on } X.$$

Now for $\alpha \geq 0$, each A_1^{α} is densely defined in X and we set

$$X_{\alpha} = D(A_1^{\alpha}).$$

For $\alpha \geq \beta \geq 0$, it can be shown that

$$X_{\alpha} \subseteq X_{\beta}.$$

Notice that $X_0 \equiv X$ and $X_1 \equiv D(A_1) \equiv D(A)$.

The A_1^{α} are called fractional powers of A_1. They satisfy $A_1^{\alpha}A_1^{\beta} = A_1^{\alpha+\beta}$. Their domains, X_{α}, provide the setting for our existence theorem.

A **solution** of the problem (1.6.4) on the time interval $(0,T)$ is a continuous function $u : [0,T) \to X \ (= L_2(\Omega))$, satisfying

$$u_t + Au = f(u), \text{ for } t \in (0,T),$$

such that

$$u(t) \to u_0 \text{ as } t \to 0 \text{ in } X, \qquad u(t) \in D(A) \text{ for } t \in (0,T).$$

Thus, the initial condition is satisfied in a general sense, while for $t > 0$, we are demanding extra smoothness which we may or may not have imposed on u_0.

Theorem. Suppose f, in (1.6.4), is a Lipschitz continuous mapping from $U = X_{\alpha}$ into X, for some $\alpha \in [0,1]$. Then, for any $u_0 \in U$, there exists $T = T(u_0)$ such that (1.6.4) has a unique solution on $(0,T)$.

The above theorem guarantees us a solution as long as u_0 is smooth enough. This restriction depends upon the nonlinearity f in (1.6.4): if f is not well-behaved, we may have to go as far as $U = D(A)$, in which case u_0 even has to satisfy the boundary conditions! This is not generally the case though. In order to apply our theory, we must do two things. Firstly we must be able to identify $D(A)$, and secondly we must develop some means by which we can ensure $f \in X$, when $u \in U = X_{\alpha}$.

Unfortunately, given general boundary conditions, it is not always possible to characterize $D(A)$. However, since $\|\Delta u\|_{L_2}$ is bounded for $u \in D(A)$, we are assured that $D(A) \subseteq H_2(\Omega)$.

Moreover, it can be shown that for domains, Ω, with smooth boundaries (e.g. when $\partial\Omega$ can be specified locally at each point, by

$$x_i = g(x_1, \ldots, x_{i-1}, x_{i+1}, \ldots, x_n),$$

where g is Lipschitz continuous [35]):

$$X_\alpha \subseteq H_k(\Omega) \;\text{ when } \alpha > \frac{k}{2}.$$

Thus, it is sufficient to show $f : U \to L_2(\Omega)$, when U is some Sobolev space $H_k(\Omega)$ (see the definitions at the back of this book).

We remark that the solution u given by the above theorem may well be arbitrarily smooth. The Sobolev embedding theorem states that if U is in $H_2(\Omega)$, then it is also in C^k where $k < 2 - \frac{n}{2}$. Thus, our solutions are certainly continuous for $n = 1, 2, 3$. In fact, there is a relatively straightforward regularity theory (see [35] Chapter 3) which allows us to show that $u \in C^\infty(\Omega)$, for $t > 0$, provided that f is *reasonable.*

Turning our attention to the nonlinearity, f, the following inequalities often prove useful when choosing U in order to apply the above theorem:

$$\|u\|_{L_p} \leq \text{constant}.\|u\|^{\theta}_{H_q} \|u\|^{1-\theta}_{L_r}, \tag{1.6.5}$$

if $0 \leq \theta \leq 1,\ p \geq q,\ p \geq r$, and

$$-\frac{n}{p} \leq \theta\left(q - \frac{n}{2}\right) - (1-\theta)\frac{n}{r},$$

with strict inequality if $r = 1$;

$$\|u\|_{C^k} \leq \text{constant}.\|u\|^{\theta}_{H_q} \|u\|^{1-\theta}_{L_r}, \tag{1.6.6}$$

if $0 \leq \theta \leq 1$, and

$$k < \theta\left(q - \frac{n}{2}\right) - (1-\theta)\frac{n}{r}.$$

These inequalities are versions of the Nirenberg-Gagliardo inequalities and hold whenever Ω is a fairly well-behaved subset of $\mathbf{R}^n$, such as described above.

We consider the following examples.

Example 1

Consider (1.6.1)-(1.6.3) where f is a polynomial in u of order p, with coefficients in $C^0(\Omega)$. Then $f \in L_2(\Omega)$ if and only if $u \in L_{2p}(\Omega)$. From (1.6.5),

$$\|u\|_{L_{2p}} \leq \|u\|_{H_q}$$

if $q \geq n(p-1)/2p$, which is trivial if $2q \geq n$.

Suppose we seek an existence theory for initial values in $U = H_1(\Omega)$. Then if $\Omega \subseteq \mathbf{R}^n$, for $n = 1,\ 2$, there is no restriction upon p, but if $n = 3$ we require $p \leq 3$.

If we go as far as asking that the initial conditions lie in $U = D(A) \subseteq H_2(\Omega)$, then there is no restriction upon p for $n = 1, \dots, 4$.

Strictly, we must also check that f is Lipschitz continuous. In this case, we proceed as follows.

Clearly for u, v in U, $f(u), f(v)$ are in $L_2(\Omega)$ and

$$\|f(u) - f(v)\|_{L_2} = O(\|u^p - v^p\|_{L_2}).$$

But,

$$\int |u^p - v^p|^2 dx = \int |u - v|^2 \left(\sum_{j=0}^{j=p-1} |v^j u^{p-1-j}| \right)^2 dx$$

$$\leq \left(\int |u - v|^{2p} dx \right)^{\frac{1}{p}} \cdot \left(\int \left(\sum_{j=0}^{j=p-1} |v^j u^{p-1-j}| \right)^{\frac{2p}{p-1}} dx \right)^{\frac{p-1}{2p}}$$

(using Hölder's inequality).

However for each $k = 0, \dots, p-1$,

$$\int |v|^{\frac{2pk}{p-1}} |u|^{\frac{2p(p-1-k)}{p-1}} dx \leq \left(\int |v|^{2p} dx \right)^{\frac{k}{p-1}} \cdot \left(\int |u|^{2p} dx \right)^{\frac{p-k-1}{p-1}}.$$

Now, since the U-norm dominates the L_{2p}-norm, we have:

$$\|f(u) - f(v)\|_{L_2}^2 \leq O\left(R^{\frac{p-1}{p}} \|u - v\|_{L_{2p}}^2 \right) = O\left(R^{\frac{p-1}{p}} \|u - v\|_U^2 \right),$$

where $\|u\|_U$, $\|v\|_U \leq R$.

Example 2

Now consider Burger's equation on $\mathbf{R}$:

$$u_t = u_{xx} - u u_x.$$

Here the nonlinear term is in $L_2(\mathbf{R})$ if and only if

$$\int_{\mathbf{R}} u^2 u_x^2 \, dx < \infty.$$

Now if $u \in U \equiv H_1(\mathbf{R})$ we have

$$\|u\|_\infty \leq \text{constant}.\|u\|_{H_1}$$

by (1.6.6). Thus

$$\int_{\mathbf{R}} u^2 u_x^2 \, dx \leq \text{constant}.\|u\|_\infty^2 \int_{\mathbf{R}} u_x^2 \, dx = \text{constant}.\|u\|_{H_1}^4.$$

Hence $f : H_1 \to L_2$.

For u, v in H_1,

$$\begin{aligned}\|f(u) - f(v)\|_{L_2} &\leq \|u(u_x - v_x)\|_{L_2} + \|(u - v)v_x\|_{L_2} \\ &\leq \|u\|_\infty \|u - v\|_{H_1} + \|u - v\|_\infty \|v\|_{H_1} \\ &\leq \text{constant}.\left(\|u\|_{H_1} + \|v\|_{H_1}\right)\|u - v\|_{H_1},\end{aligned}$$

(by (1.6.6)), which shows that $f : H_1 \to L_2$ is Lipschitz on bounded subsets.

Notice that if we try to choose U any larger than H_1 we could not be assured of u_x in L_2, let alone uu_x!

Example 3

Consider (1.6.1)-(1.6.3) with $f = \sum_{j=1}^n a_j(x)u_{x_j}$, where the a_j are uniformly bounded (in BC, for example). In this case we incorporate all such terms into the definition of the operator A, which will remain sectorial. That is, for functions u in $L_2(\Omega)$ with Δu in $L_2(\Omega)$, which satisfy any boundary conditions, (1.6.2), we define

$$Au = -\Delta u - \sum_{j=1}^n a_j(x)u_{x_j},$$

and extend it to become a closed, densely defined operator in $X = L_2(\Omega)$ as usual.

Example 4

Let $E(u)$ be a polynomial of order m, $\Omega \equiv \mathbf{R}^n$ and $k(x) \in L_p(\mathbf{R}^n)$. Then consider (1.6.1), (1.6.2) where

$$f = k * E(u) = \int_{\mathbf{R}^n} k(x - y)E(u(y)) \, dy.$$

Now Young's inequality (see Box D or [22]) ensures that

$$\|f\|_{L_2} \leq \|k\|_{L_p}.\|E(u)\|_{L_q},$$

where $\frac{1}{p}+\frac{1}{q}=\frac{3}{2}$, provided that the norms on the right-hand side are finite. Clearly $E(u) \in L_q$ if and only if $u \in L_{qm}$. Now if $u \in H_1$, (1.6.5) implies $u \in L_r$ where

$$-\frac{n}{r} \leq 1-\frac{n}{2} \quad \text{and } r \leq 2.$$

Thus, for $n = 1, 2, 3$ we must have $r = qm \leq 2$, in order that $f \in L_2$. Thus, $f : H_1 \to L_2$ so long as

$$\frac{2mp}{3p-2} \leq 2.$$

Box F: Analytic semigroups

Here, we sketch some of the considerations that lie behind the local existence theorem quoted in this section. We use a general approach, and the first-time reader is advised to think of the Banach space, X, introduced below, as $L_2(\mathbf{R})$, and the linear operator, A, as being induced on X by $-\Delta$, or something similar.

A linear operator, A, in a Banach space, X, is called **sectorial** if it is closed, densely defined, and such that there are constants, ϕ, in $(0, \pi/2)$, $M \geq 0$, and $a \in \mathbf{R}$ such that

$$R = \{\lambda \in |C| \phi \leq |\arg(\lambda - a)| \leq \pi, \quad \lambda \neq a\}$$

contains no part of the spectrum of A, and that the estimate

$$\|(\lambda - A)^{-1}\| \leq M/|\lambda - a|$$

holds in R.

An **analytic semigroup** on X is a family of continuous linear operators on X, $\{S(t)\}_{t\geq 0}$, satisfying
(i) $S(0) = I$, the identity; and

$$S(t)S(s) = S(t+s), \quad t \geq 0, \ s \geq 0,$$

the semigroup property;
(ii) $S(t)x \to x$ as $t \to 0^+$ for all $x \in X$;
(iii) The mapping $t \to S(t)x$ is analytic on $0 < t < \infty$ for all $x \in X$.

The **infinitesimal generator**, B, of this semigroup is defined by

$$Bx = \lim_{t\to 0^+} \frac{1}{t}(S(t)x - x)$$

and has domain containing all those $x \in X$ for which the limit exists in X. We exploit the analogy with finite-dimensional spaces (for example, $X = \mathbf{R}^n$) to write

$$S(t) = e^{Bt}.$$

(Note that if $X = \mathbf{R}$, the semigroup property may be solved to give $S(t) = e^{\alpha t}$ for any constant $\alpha \in \mathbf{R}$.)

Here is the connection. If A is a sectional operator, then $-A$ is the infinitesimal generator of an analytic semigroup. Moreover,

$$\frac{d}{dt}e^{-At}x_0 = -Ae^{-At}x_0, \quad \text{for all } x_0 \in X.$$

Thus, $x = e^{-At}x_0$ solves the general initial-value problem

$$\begin{aligned} x_t + Ax &= 0 \quad t > 0, \\ x(0) &= x_0. \end{aligned}$$

Here, by a solution on a time interval $(0, T)$, we mean a continuous function $x : [0, T) \to X$, continuously differentiable with $x \in D(A)$ for $t \in (0, T)$, satisfying $x_t + Ax = 0$, and $x(t) \to x_0$ in X as $t \to 0$.

Once this linear problem is solved, the resulting semigroup is utilized in order to solve the nonlinear problem

$$\begin{aligned} u_t + Au &= f(u) \quad t > 0, \\ u(0) &= u_0. \end{aligned}$$

Under *reasonable* conditions upon f and u_0, the *variation of parameters* formula holds (algebraically, we treat e^{At} in much the same fashion as a scalar exponential function). We refer to [35], [38], for a detailed construction and proof.

The good thing about semigroups is that many dynamic processes can be thought of in these terms, partial differential and ordinary differential equations alike.

Note that we only require $S(t)$ to be defined for $t > 0$. If $S(t)$ is also defined for all negative t, then the semigroup property implies $S(-t) = S(t)^{-1}$. In this case, our semigroup is, in fact, a **group**.

Example 1

Consider the scalar ordinary differential equation

$$y_t = \alpha y \quad t > 0.$$

Here, $S(t) : \mathbf{R} \to \mathbf{R}$, for each t, is defined by

$$S(t)y = e^{\alpha t}y.$$

In this case, $S(t)_{t\in\mathbf{R}}$ is a group.

Example 2

Consider the diffusion equation in one dimension

$$u_t = u_{xx}, \quad x \in \mathbf{R},\ t > 0.$$

In $L_2(\mathbf{R})$, we have $S(t) : L_2(\mathbf{R}) \to L_2(\mathbf{R})$, defined explicitly in this case by

$$S(t)w(x) = \int_{-\infty}^{\infty} G(x-y,t)w(y)dy$$

where G is the fundamental solution,

$$G(x,t) = \frac{1}{\sqrt{4\pi t}} \exp\frac{-x^2}{4t}.$$

Notice that G, and hence S, is not defined for negative t.

There is an excellent introductory account of semigroups given in [46]. Those interested in applications to partial differential equations are referred to [38],[63]. Pazy's notes are the standard reference.

1.7 Blow-up

We have seen that solutions of reaction-diffusion equations exist locally in time, provided that the initial data is smooth enough. In many cases, the solutions exist for all subsequent time (i.e. globally). In other cases, solutions exist for only a finite time. We associate this behaviour with one or more of the state variables becoming large – large enough, that is, to escape from the class of functions where (local) existence is ensured. In examples, we generally find that solutions develop a singularity at a point, or points, within the domain. The singularity may be a point where the dependent variable itself reaches infinity, or it may be a point where a discontinuity (or shock) develops. Either way, if this happens in finite time and the solution cannot be continued, we refer to this phenomenon as **blow-up**.

In combustion theory, it is usually the temperature that blows up. Such behaviour is often termed *thermal runaway* – a less alarming, but equally graphic, description.

Consider the following example:

$$u_t = u_{xx} - u + u^p, \quad x \in (0,\pi), \quad t > 0, \tag{1.7.1}$$

$$u(x,0) = u_0(x) \geq 0, \quad x \in (0,\pi),$$

$$u(0,t) = u(\pi,t) = 0, \quad t > 0.$$

Here, p is some fixed power and local existence is guaranteed for all $u_0 \in H_1(0,\pi)$. It is easy to show (using comparison principles, for example – see section 1.8) that $u_0(x) \geq 0$ implies $u(x,t) \geq 0$ wherever it exists.

Let

$$f(t) = \int_0^\pi u(x,t)\sin x \, dx.$$

Multiplying (1.7.1) by $\sin x$, and integrating over $(0,\pi)$, we obtain

$$\begin{aligned} f_t &= \int_0^\pi u_{xx}\sin x \, dx - f + \int_0^\pi u^p \sin x \, dx \\ &= -2f + \int_0^\pi u^p \sin x \, dx. \end{aligned}$$

Here we used integration by parts twice on the first term.

Now using Hölder's inequality:

$$\int_0^\pi u \sin x \, dx \leq \left(\int_0^\pi u^p \sin x \, dx\right)^{\frac{1}{p}} . \left(\int_0^\pi \sin x \, dx\right)^{1-\frac{1}{p}}.$$

Hence

$$f_t \geq -2f + \frac{f^p}{2^{p-1}}.$$

So if

$$f(0) = \int_0^\pi u_0(x)\sin x \, dx > 2^{\frac{p}{p-1}},$$

f tends to ∞ in finite time, and certainly before

$$t = \frac{1}{2^{p-1}} \int_{f(0)}^\infty \frac{df}{(f^p - 2^p f)}.$$

Now, by the Cauchy-Schwartz inequality

$$f \leq \|u\|_{L_2(0,\pi)} . \|\sin x\|_{L_2(0,\pi)}.$$

Thus, $f \to \infty$ implies $\|u\|_{L_2(0,\pi)} \to \infty$. So if $f(0) > 2^{p/(p-1)}$, then u will leave $L_2(0,\pi)$ in finite time.

1.8 Comparison principles

One of the most well-known qualitative techniques in reaction-diffusion is that of comparison principles. If a reaction-diffusion system allows their use, then many problems may be attacked: existence of solutions, existence of travelling waves, even delicate stability phenomena. The catch is that not all systems are amenable to such an approach.

We begin by considering a scalar equation

$$u_t = \Delta u + f(u, \mathbf{x}, t), \quad \mathbf{x} \in \Omega \subseteq \mathbf{R}^n, \ t > 0.$$

Here f is smooth (say, continuously differentiable).

Suppose $\overline{u}$ satisfies $\overline{u} : \Omega \times [0, T] \to B$, some bounded subset of $\mathbf{R}$, and

$$\overline{u}_t \geq \Delta \overline{u} + f(\overline{u}, \mathbf{x}, t);$$

then we say that $\overline{u}$ is a **super-solution**. If $\underline{u}$ satisfies $\underline{u} : \Omega \times [0, T] \to B$, and

$$\underline{u}_t \leq \Delta \underline{u} + f(\underline{u}, \mathbf{x}, t),$$

then we say that $\underline{u}$ is a **sub-solution**.

Now suppose that there exist constants α, β ($\alpha^2 + \beta^2 \neq 0$) such that

$$\alpha\overline{u} - \beta\nabla\overline{u}.\mathbf{n} \geq \alpha\underline{u} - \beta\nabla\underline{u}.\mathbf{n}, \quad \mathbf{x} \in \partial\Omega, \ t > 0,$$

($\mathbf{n}$ is the outer normal to $\partial\Omega$); and that

$$\overline{u}(\mathbf{x}, 0) \geq \underline{u}(\mathbf{x}, 0) \quad \mathbf{x} \in \Omega.$$

Then we claim that

$$\overline{u}(\mathbf{x}, t) \geq \underline{u}(\mathbf{x}, t) \quad \mathbf{x} \in \Omega, \ t > 0.$$

To see this, set $w = \overline{u} - \underline{u}$. Then the mean value theorem implies

$$w_t - \Delta w \geq f_u(\underline{u} + \theta(\overline{u} - \underline{u}))w$$

for some mapping $\theta : \Omega \times [0, T] \to [0, 1]$. The result follows from the strong maximum principle for linear parabolic equations [18]. (The point is that w is initially nonnegative and the boundary conditions ensure that if it becomes negative it must do so at an interior point of Ω. We can then obtain a contradiction.)

This is our comparison principle. To use it, we try to bound our unknown solution above and below by super- and sub- solutions respectively.

Note that if the super- and sub- solutions remain bounded for all time t, then we have infinity bounds for our solution **wherever** it exists. Thus, on bounded domains, we obtain a priori estimates which allow us to pass from local to global existence.

In order to apply a comparison principle to systems, we consider the equation

$$\mathbf{u}_t = D\Delta\mathbf{u} + \mathbf{f}(\mathbf{x}, t, \mathbf{u}), \quad \mathbf{x} \in \Omega,\ t > 0, \tag{1.8.1}$$

where

$$\mathbf{u} = (u_1, \ldots, u_m)^T,$$
$$\mathbf{f} = (f_1, \ldots, f_m)^T,$$

with $f_i = f_i(\mathbf{x}, t, \mathbf{u})$, and

$$D = \mathrm{diag}(d_1, \ldots, d_m),$$

where $d_j \geq 0$.

We define the vector inequality $\mathbf{v} \geq \mathbf{w}$ in $\mathbf{R}^m$ to mean $v_i \geq w_i$ for each $i = 1, \ldots, m$.

Proceeding as before, setting $\mathbf{w} = \overline{\mathbf{u}} - \underline{\mathbf{u}} \in \mathbf{R}^m$, we must find a constant, Q_i (for each $i = 1, \ldots, m$), such that

$$f_i(\overline{\mathbf{u}}, \mathbf{x}, t) - f_i(\underline{\mathbf{u}}, \mathbf{x}, t) \geq (\overline{u}_i - \underline{u}_i)Q_i.$$

Here we allow $\overline{\mathbf{u}}$ and $\underline{\mathbf{u}}$ to range over some compact subset, B, of $\mathbf{R}^m$, whilst maintaining $\overline{\mathbf{u}} - \underline{\mathbf{u}} \geq 0$.

Clearly, such a condition cannot hold if f_i is decreasing in B with respect to any of the variables u_j $(j \neq i)$. Conversely, if the derivatives $f_{i\ u_j}(\mathbf{u}, x, t)$ $(j \neq i)$ are nonnegative for $\mathbf{u} \in B$, $\mathbf{x} \in \Omega$, $t > 0$, then constant Q_i may be found. In this case (1.8.1) will admit a comparison principle, relating sub- and super-solutions, $\underline{\mathbf{u}}$, $\overline{\mathbf{u}}$, taking values in B.

Example

Consider the scalar equation

$$u_t = \Delta u + u(1-u)(u-a), \quad \mathbf{x} \in \Omega,\ t > 0,$$

(a is a constant in (0,1)), Ω a bounded set in $\mathbf{R}^n$, together with no-flux boundary conditions

$$\nabla u.\mathbf{n} = 0, \quad \mathbf{x} \in \partial\Omega,\ t > 0.$$

If $0 \leq u(\mathbf{x}, 0) \leq 1$ in Ω, then the comparison principle implies that $0 \leq u(\mathbf{x}, t) \leq 1$ wherever the solution exists (take $\overline{u} \equiv 1$, $\underline{u} \equiv u$, etc.). Since this is an *a priori* bound, we can extend our solution to be global (i.e. exist for all $t > 0$).

If $0 \leq u(\mathbf{x}, 0) < a$ in Ω, then the comparison principle implies $u \to 0$ as $t \to \infty$, faster than $s(t)$, where

$$s_t = s(1-s)(s-a), \quad s(0) = \max_{\mathbf{x}\in\Omega} u(x, 0)$$

(take $\overline{u} \equiv s$, $\underline{u} \equiv u$).

We shall refer to comparison principles in some specific examples later. The main point to remember is that when the nonlinearities satisfy the above monotonicity condition, a comparison principle is probably the most useful qualitative tool to use in a variety of settings. The interested reader is referred to Fife's book [18], as well as [3],[20],[30],[73] amongst others.

1.9 Invariant regions

A qualitative technique related to comparison principles is that of positive invariant regions [9],[29],[65],[70]. The idea is to find a (hopefully compact) set in the state space from which solutions cannot escape. Thus, we may go on to obtain *a priori* bounds and global existence, as well as information about the long-time behaviour of solutions.

Consider the system

$$\mathbf{u}_t = D\Delta\mathbf{u} + \mathbf{f}(\mathbf{u}), \quad \mathbf{x} \in \Omega,\ t > 0.$$

Here $\mathbf{u} \in \mathbf{R}^m$, $\mathbf{f} : \mathbf{R}^m \to \mathbf{R}^m$ is some continuous vector field, and $D = \text{diag}(d_1, \ldots, d_m)$, where each $d_i \geq 0$ is the diffusion matrix.

Let B be a block of the form

$$B = \prod_{j=1}^{m} [\underline{u}_j, \overline{u}_j] \tag{1.9.1}$$

for constants $\underline{u}_j < \overline{u}_j$. Let $\mathbf{N}$ denote any outer normal to B on ∂B. Suppose that

$$\mathbf{f}.\mathbf{N} < 0, \quad \mathbf{u} \in \partial B.$$

Then we claim that any solution $\mathbf{u}$ with $\mathbf{u}(\mathbf{x}, 0)$ inside B for all $\mathbf{x} \in \Omega$, and boundary values inside B, must remain inside B wherever it exists.

We may also allow any boundary conditions that satisfy

$$\nabla u_i.\mathbf{n} < 0 \text{ if } u_i = \overline{u}_i,$$

and

$$\nabla u_i.\mathbf{n} > 0 \text{ if } u_i = \underline{u}_i,$$

where $\mathbf{n}$ is the outer normal on $\partial\Omega$, as usual. The point is that, for such conditions, if $\mathbf{u}$ leaves B, it must do so at some point in the interior of Ω: say $\mathbf{x} = \mathbf{x}^*$, $t = t^*$ (check this for yourself).

Suppose, without loss of generality, that

$$u_i(\mathbf{x}^*, t^*) = \overline{u}_i$$

and $\mathbf{u} \in \text{int}B$ for all $\mathbf{x} \in \Omega, 0 < t < t^*$. Then

$$u_{i\ t}(\mathbf{x}^*, t^*) \geq 0.$$

However, $u_i(., t^*)$ must have a maximum at $\mathbf{x}^*$. Thus

$$\begin{aligned} u_{i\ t} =& d_i \Delta u_i + f_i(\mathbf{u}) \\ \leq& f_i(\mathbf{u}). \end{aligned}$$

But, at $u_i = \overline{u}_i$, we chose the outer normal, $\mathbf{N}$, to be the ith unit vector. So

$$\mathbf{f}.\mathbf{N} = f_i(\mathbf{u}) < 0$$

by hypothesis. Thus $u_{i\ t} < 0$, which is a contradiction. Hence $\mathbf{u}(\mathbf{x}, t)$ must remain inside B.

The requirement that B be an oriented block, of the form (1.9.1), can be relaxed only if two or more of the diffusion coefficients are equal. The general condition [9] demands that the block B is such that all outer normals, $\mathbf{N}$, are left eigenvectors of the diffusion matrix, D (introduce any change of state variables which makes D diagonal).

Example The FitzHugh-Nagumo equations.

These equations are a simple model for the conduction of action potentials along unmyelinated nerve fibres, see [21],[55]. It is often utilized by physiologists and mathematicians because it maintains the qualitative behaviour of solutions, without introducing the complexity of more realistic models such as the Hodgkin-Huxley equations, [36].

We have

$$\begin{aligned} u_t =& \ u_{xx} + f(u) - v \\ v_t =& \ \delta v_{xx} + \sigma u - \gamma v, \end{aligned} \tag{1.9.2}$$

where σ, γ are positive; $\delta \geq 0$; and f has the qualitative form $u(1-u)(u-a)$. We choose B to be the rectangle depicted in Figure 1.11. The nullclines of the reaction terms are shown and it is easy to see that B is positively invariant for (1.9.2).

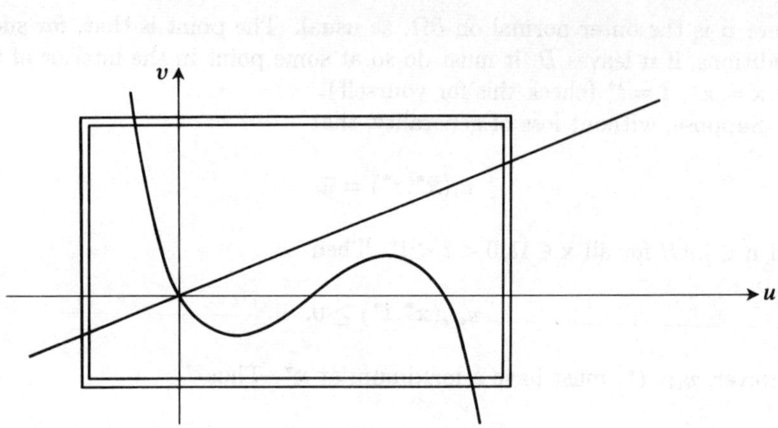

Figure 1.11: The nullclines for the nonlinearities in (1.9.2)

Example Competition Equations in Population Biology.

We consider two competing populations which interact according to the Lotka-Volterra competition rules. Allowing for the random motion of individuals, we have

$$\begin{aligned} u_t &= d_1 u_{xx} + u(A_1 - B_1 u - C_1 v) \\ v_t &= d_2 v_{xx} + v(A_2 - B_2 u - C_2 v). \end{aligned}$$

Here, A_i, B_i, and C_i are positive constants, while the diffusion coefficients d_i are nonnegative. Depending upon the choices of A_i, B_i, and C_i, we may have three or four rest points. Figures 1.12 (a)-(d) depicts the various configurations.

We check that the set

$$B = \{(u, v) : \ 0 \leq u \leq A_1/B_1 + \varepsilon, \ 0 \leq v \leq A_2/C_2 + \varepsilon\}$$

is always positively invariant, for any $\varepsilon > 0$. Along the upper edges, the usual invariant region argument, given above, applies, whilst to show $u \geq 0, v \geq 0$, we may use the estimates

$$u_t \geq d_1 u_{xx} - K_1 u$$

$$v_t \geq d_2 v_{xx} - K_2 v$$

where $K_i = \max_B(-A_i + B_i u + C_i v)$. The usual strong maximum principle (or comparison principle) applied to

$$w_t = d_i w_{xx} - K_i w$$

shows that the super-solutions u and v stay positive in B if they start out there.

Along with B, we also depict some other positively invariant blocks in Figure 1.12 (a)-(d). One may easily check their invariance.

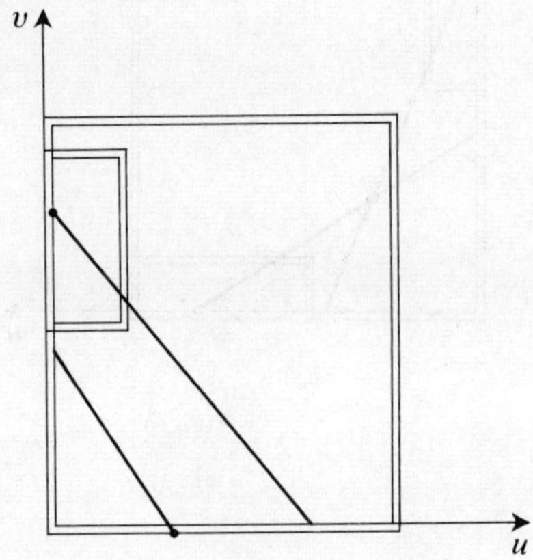

(a)

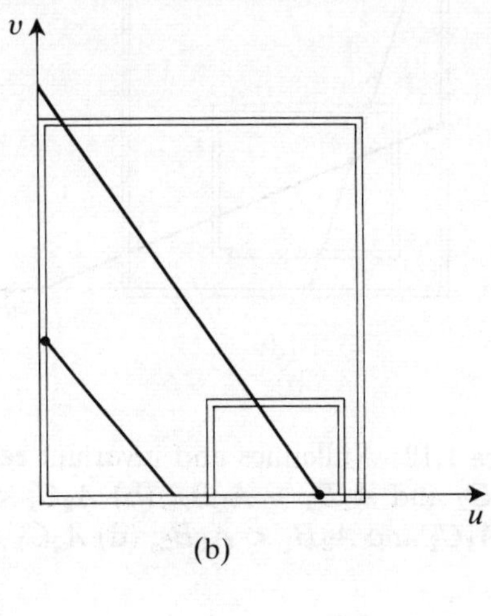

(b)

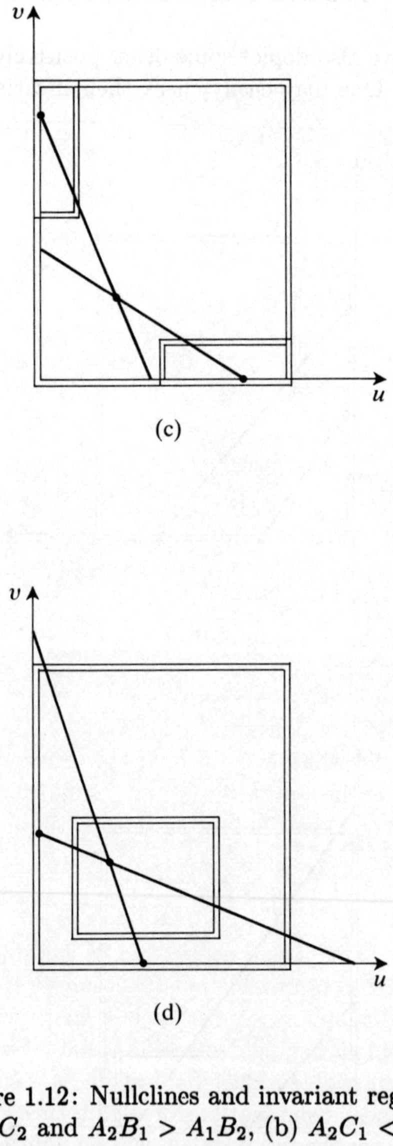

Figure 1.12: Nullclines and invariant regions.

(a) $A_2C_1 > A_1C_2$ and $A_2B_1 > A_1B_2$, (b) $A_2C_1 < A_1C_2$ and $A_2B_1 < A_1B_2$, (c) $A_2C_1 > A_1C_2$ and $A_2B_1 < A_1B_2$, (d) $A_2C_1 < A_1C_2$ and $A_2B_1 > A_1B_2$.

Exercise 1.1 Hydrodynamic dispersion

Consider the problem of the diffusion and advection of a soluble tracer in a water-filled capillary, given in cylindrical coordinates by $r < a$. The fluid flow along the tube is *slow flow* (see Box I) in the axial z-direction and is given by

$$u(r, z) = 2\overline{u}(1 - \frac{r^2}{a^2}),$$

where $\overline{u}$ is the cross-sectional average flow rate.

Assuming that the tracer is subject to no-flux boundary conditions at the capillary surface ($r = a$), use the balance law to show that the tracer concentration $c(r, z, t)$ satisfies

$$c_t = D_m(c_{zz} + r^{-1}(rc_r)_r) - (u(r, z)c)_z, \quad 0 < r < a,$$

$$c_r = 0 \text{ at } r = a,$$

where D_m denotes the molecular diffusivity of the tracer in water.

Let $\overline{c}(z, t)$ denote the cross-sectional average tracer concentration:

$$\overline{c} = \frac{2}{a^2}\int_0^a c(r, z, t) r\, dr.$$

Suppose that we wish to simplify the above problem for c to obtain a one-dimensional linear model for advection and dispersal in the axial z-direction alone in the form

$$\tilde{c}_t = D^* \tilde{c}_{zz} - u^* \tilde{c}_z - s^* \tilde{c}, \quad z \in \mathbf{R},\ t > 0,$$

where D^*, u^*, and s^* are constants. We would hope that the solution $\tilde{c}$ could be close to $\overline{c}(z, t)$, which is defined via the solution to the full problem. How should we choose the constants D^*, u^*, and s^* for this to be the case?

a) Seek solutions $c(r, z, t)$ to the full problem in the form

$$c = f(r)e^{\sigma t + ikz}$$

for wave numbers k. This is just separation of variables. Derive a Sturm-Liouville problem for $f(r)$, which must be solved for the eigenvalues σ as well as f depending upon k. For each k there will be a sequence of solutions to this problem having decreasing values of σ asymptotic to $-\infty$. A solution to the full problem may be written as a sum or integral (with respect to k) over linear combinations of such terms.

Any solution, c, for the full problem will be dominated by those terms having $Re(\sigma)$ relatively to the right of the complex plane.

b) Expand σ and f in powers of k:

$$\sigma = \sigma_0 + k\sigma_1 + k^2\sigma_2 + O(k^3),$$

$$f = f_0 + kf_1 + k^2 f_2 + O(k^3),$$

where each f_n satisfies the boundary conditions

$$\frac{df_n}{dr}(a) = 0, \quad f \text{ bounded at } 0.$$

Substitute these series into the Sturm-Liouville problem and equate coefficients.

Consider the zeroth-order equation for σ_0 and f_0. Choosing $e^{Re(\sigma_0)}$ to be as large as possible (so as to identify the dominant terms), show that $\sigma_0 = 0$ and $f_0(r) \equiv 1$.

Consider the $O(k)$ equation. Show that we must choose $\sigma_1 = -i\overline{u}$ (a solvability criterion), and obtain the subsequent solution

$$f_1(r) = \frac{i2\overline{u}}{16D_m}(2r^2 - \frac{r^4}{a^2}).$$

Next consider the $O(k^2)$ equation. Show that we must set

$$\sigma_2 = -D_m - \frac{a^2\overline{u}^2}{48D_m}$$

in order that a solution exists.

Thus solutions c of the full problem are dominated by terms of the form

$$e^{ikz+\sigma(k)t}(1 + k\frac{i2\overline{u}}{16D_m}(2r^2 - \frac{r^4}{a^2}) + O(k^2)),$$

where

$$\sigma(k) = -ik\overline{u} - k^2(D_m + \frac{a^2\overline{u}^2}{48D_m}) + O(k^3).$$

c) Find the dispersion curve for the one-dimensional equation to be solved for $\tilde{c}$ (substitute $\tilde{c} = \exp ikz + \sigma(k)t$).

Choosing

$$s^* = \sigma_0 = 0, \ u^* = \overline{u}, \text{ and } D^* = D_m + \frac{a^2\overline{u}^2}{48D_m},$$

the spectra of the problems are asymptotically matched for small k up to second order.

Note that $D^* > D_m$. The additional part of the diffusivity in the one-dimensional, axial, problem is called Taylor dispersion. It accounts for the ability of the tracer to diffuse across the profile of the flow rate, given by $u(r)$, giving rise to dispersion due to the variability of the advection term. The approximation of the full problem by the above one-dimensional problem is valid in the limit of large t, or any initial value problem where the behaviour is dominated by small wave numbers (relatively smooth solutions).

Exercise 1.2

If we apply the Dirichlet condition, $c = 0$ at $r = a$, instead of the no-flux condition, then there will be a net outflux of tracer at the capillary surface. Following the steps in the previous question, show that we must choose $\sigma_0 \neq 0$ in this case. Find the values for $s^*(= \sigma_0)$, $u^*(= i\sigma_1)$, and $D^*(= -\sigma_2)$ for this problem. (Hence, the level of Taylor dispersion in axial transport equations depends upon the boundary conditions imposed on the full problem.)

Exercise 1.3

Consider the approximation of the behaviour of $\bar{c}$ by that of $\tilde{c}$ in the previous questions. For readers familiar with the method of steepest descent [52], show that we have

$$|\bar{c}(z,t) - \tilde{c}(z,t)| \leq O(e^{-s^* t}/t)$$

as $t \to \infty$, provided that $\tilde{c}(z, 0)$ is chosen appropriately.

This result clarifies the sense in which $\tilde{c} \approx \bar{c}$, when the Taylor dispersion coefficient, together u^* and s^*, are chosen appropriately.

Exercise 1.4

Generalize the results in the previous three questions so that

a) c satisfies

$$\alpha c_r + \beta c = 0 \quad \text{at} \quad r = a \quad (\alpha, \beta \geq 0).$$

b) Cross-fracture advection is included at a given rate $v(r)$ in the radial direction, but there is no outflux of particles at the boundaries.

Exercise 1.5

A population of cars on a one-dimensional road travel in the positive x-direction with speed $1 - u$, where $u(x,t)$ is the car density ($u = 1$ corresponding to the *nose to tail* maximum congestion and $u = 0$ corresponding to an empty road). Using the basic balance law, show that u must satisfy

$$u_t + [u(1-u)]_x = 0$$

provided that there are no source or sink terms (sliproads!).

Given $u(x,0) = u_0(x)$, a differentiable function from $\mathbf{R}$ into $[0,1]$, show that a smooth solution exists as far out as

$$t = \inf_{s\in\mathbf{R}} \left\{ \frac{1}{2u_0'(s)} \right\}$$

where $'$ denotes $\frac{d}{ds}$.

At this time, separate characteristics converge and a **shock** develops (see Chapter 3). What is the effect upon individual motorists? Show that if $u_0 \approx$ constant, then any perturbations are propagated in such a way that negative values of u_x are increasing along characteristics $u_{xt} = \frac{2u'^2}{(1-2tu')^2}$. Thus, any increasing profiles in the initial density distributions become steeper as the solution evolves (along characteristics), whilst decreasing profiles become flatter.

This illustrates why, in traffic jams on motorways, drivers find themselves subject to a stop-go regime rather than all cars moving at some uniform speed.

Exercise 1.6

We investigate the travelling wave solutions for the following problem:

$$\begin{aligned} u_t &= u_{xx} + kvg(u) \\ v_t &= -\, vg(u), \end{aligned}$$

Here k is some positive constant and g is a monotone increasing function satisfying $g(0) = 0$.

In this model, u represents the temperature distribution, while v denotes the concentration of some immobile chemical species, which is combustible at positive temperatures. Think of a burning fuse.

We shall impose the boundary conditions

$$\begin{aligned} (u,v) &\to (0,\tilde{v}) \text{ as } x \to +\infty \\ (u,v) &\to (\tilde{u},0) \text{ as } x \to -\infty \end{aligned}$$

for some suitable nonnegative constants $\tilde{u}$ and $\tilde{v}$. Thus, to the right the chemical (fuel) is intact, whilst to the left it has been consumed. Let us look for a travelling wave solution.

Introducing the moving frame $z = x - ct$ and writing $u = u(z)$ etc., we have

$$\begin{aligned} u_{zz} + cu_z + kg(u)v &= 0 \\ cv_z - g(u)v &= 0. \end{aligned}$$

Show that this system may possess a solution commensurate with the boundary conditions only if

$$\tilde{u} = k\tilde{v}.$$

Thus, we may define $\tilde{u}$ to be consistent with the given value of $\tilde{v}$. This being so, write the resulting problem as

$$u_{zz} + cu_z + g(u)\left[(\tilde{u} - u) - \frac{u_z}{c}\right] = 0.$$

Let us assume $g(u) = g_0 u$ for some possible constant g_0. Show that the rest point $(\tilde{u}, 0)$ is a saddle point, in the (u_1, u_2)-phase plane, for all c.

Let T denote the triangle given by

$$\begin{aligned} 0 &\leq u \leq \tilde{u} \\ -ru &\leq u_z \leq 0. \end{aligned}$$

Show that no orbits may leave T provided

$$0 \leq (cr - r^2 - g\tilde{u}) + gu(1 - r/c)$$

for all $u \in [0, \tilde{u}]$. Hence, show that there exists a travelling wave solution such that

$$\begin{aligned} u_z &\to 0 \text{ as } z \to \infty \\ u_z &\to \tilde{u} \text{ as } z \to -\infty \end{aligned}$$

for all values of c greater than $2\sqrt{g\tilde{u}}$.

Exercise 1.7

Generalize the previous result (i) for any montonically increasing function g such that $g'(0) \neq 0$; and (ii) for the case where $g \equiv 0$ for $u \in [0, \overline{u}]$, and $g = g_0(u - \overline{u})$ for $u > \overline{u}$, where $\overline{u}$ is constant.

This last case is the most realistic, $\overline{u}$ being the ignition temperature of the fuel or fuse.

Exercise 1.8

In section 1.4 we considered the problem

$$u_{yy} + L^2 u(1 - u) = 0, \quad y \in (0, 1); \quad u = 0, \quad y = 0, 1,$$

seeking a bifurcation from the trivial solution, $u = 0$ and $L = \pi$.

We introduced a small parameter ε and set $L = \pi + \varepsilon$, and expanded u as an asymptotic series in powers of ε:

$$u = \varepsilon u_1 + \varepsilon^2 u_2 + \varepsilon^3 u_3 + \cdots.$$

To order ε we obtained the solution

$$u_1 = A \sin \pi y,$$

where A was some constant still to be determined.

To order ε^2:

$$u_{2\,yy} + \pi^2 u_2 = \pi^2 u_1^2 - 2\pi u_1, \quad u_2(0) = u_2(1) = 0.$$

We showed that $A = 3/4$.

Find $u_2(y)$ from the above inhomogeneous equation (Box C may be helpful), and by considering the terms of order ε^3, obtain an equation for u_3.

2 Pattern formation

2.1 Introduction

In many of the applications of reaction-diffusion systems, it is the long-term behaviour of solutions that is important. Here, we follow on from section 1.4 and investigate equilibrium solutions and their properties. Moreover, we shall focus upon equilibria which display nontrivial **pattern** (i.e., inhomogeneous spatial structure).

For example, in the process of **morphogenesis**, embryonic cells must somehow order themselves so as to lay down the differentiated structures (organs, limbs, etc.) which constitute a recognizable individual. One theory put forward to explain this feat is that the development follows a kind of map, or prepattern, laid down within the growing cell mass by chemicals called morphogens. These are assumed to react and diffuse throughout the medium. Clearly, the theory rests upon the ability of the resulting model to exhibit stable solutions with a high degree of pattern. In this chapter, we shall meet such systems, and shall indicate techniques by which their solutions may be analysed. Returning to morphogenesis for the moment, it is worth remarking that the morphogen prepattern theory is by no means universally accepted by theoretical biologists. The problem is that no real morphogens have been identified, so the models remain phenomenological. Recently, some new theories have been proposed which extend the old reaction-diffusion models. However, the analytical methods developed for reaction-diffusion models continue to be of use in the investigation of these alternative models of morphogenesis. For example, the models of [53], [54], are based on the coupling of mass conservation laws (like those of section 1.2) with mechanical considerations which reflect the forces that are at work within the extracellular medium. The resulting mechano-chemical models couple reaction-diffusion processes with force balance equations. We shall consider one such system in section 2.4.

Elsewhere, in **population biology** there has long been an interest in patterned equilibria. The idea that individuals of different species separate out from each other as a result of interspecific competition has led to many interesting investigations. Reaction-diffusion has played a leading role here (see [45],[57],[68] and the references therein).

Here again, analytical techniques are required to be capable of investigating systems of coupled reaction-diffusion equations. In the present chapter, we follow two distinct approaches. The first is concerned with the bifurcation of stable patterned equilibria from trivial uniform equilibria. The essence of such results is that they are local. They allow us to obtain a lot of information about the structure of the bifurcating equilibria, but only near to the bifurcation point.

In order to analyse equilibria which are far away from uniformity, a different strategy must be adopted. Here, we are sometimes faced with large systems, and cannot hope to get anywhere by linearization or regular expansions. For some systems, help is at hand in the form of singular perturbation theory (in this context, we mean the method of matched asymptotic expansions). The idea is to exploit the smallness (or largeness) of certain constants or parameters, and solve reduced systems by taking limits formally. The resulting solution segments are then *matched* together in order to obtain asymptotic representations of the original problem. The patchwork is achieved by introducing **interfaces**, or **transition layers**, in which one or more variables may move sharply from one state to another. We shall introduce this directly in section 2.3.

The idea that homogeneous, unforced systems of partial differential equations can exhibit heterogeneous space-dependent structures dates back to Turing's work, [75], in the 1950s. Section 2.2 below includes a discussion of diffusion-driven instability, which is usually called **Turing instability**.

Throughout this chapter, we restrict our consideration to systems defined on bounded domains, Ω, with **no-flux** conditions applied on the boundary, $\partial\Omega$.

Such systems are called **closed** systems since the no-flux condition allows no external influence upon the resulting solutions. It is easy to imagine patterns developing if, say, inhomogeneous Dirichlet boundary conditions were applied. In this case, the structures would be **flux-driven**, by the inflow or outflow of *particle* mass at the boundaries. Many of the analytical methods we develop can be applied to such problems, so there is no real loss in restricting our attention to closed systems. Moreover, since the no-flux (or Neumann) conditions are so passive, it is reasonable to state that closed systems exhibiting patterned equilibria are doing so with the least encouragement. Other boundary conditions, representing external forcing, are likely to enhance and develop the structures further.

2.2 Turing instability and local bifurcation

Consider the following system for the state variable $\mathbf{w} \in \mathbf{R}^2$;

$$\begin{aligned} \mathbf{w}_t &= D\Delta\mathbf{w} + \mathbf{f}(\mathbf{w}) \quad \mathbf{x} \in \Omega,\ t > 0, \\ \nabla w_i.\mathbf{n} &= 0, \quad \mathbf{x} \in \partial\Omega,\ t > 0,\ i = 1, 2. \end{aligned} \tag{2.2.1}$$

Here, D is the positive definite diagonal matrix $\mathrm{diag}(d_1, d_2)$ and $\mathbf{f} : \mathbf{R}^2 \to \mathbf{R}^2$ is some smooth vector field.

Now suppose that $\mathbf{w} \equiv \mathbf{w}_0$ is a spatial, homogeneous equilibrium for (2.2.1) (i.e.; $\mathbf{f}(\mathbf{w}_0) = 0$), then we ask whether any inhomogeneous equilibria may bifurcate from $\mathbf{w}_0$ as the diffusivities d_1 and d_2 are varied, or alternatively, as we allow the domain, Ω, to change continuously (see later remarks).

Of particular interest is the case where $\mathbf{w}_0$ is always **stable** with respect to spatially homogeneous perturbations (since otherwise arbitrary small perturbations would always grow and the solution evolve towards some other long-time regime – clearly, in this case, $\mathbf{w} = \mathbf{w}_0$ would be of limited interest in any application). Thus, we assume henceforth that $\mathbf{w}_0$ is a stable solution of the ordinary differential system,

$$\mathbf{w}_t = \mathbf{f}(\mathbf{w}), \tag{2.2.2}$$

associated with (2.2.1). Trivially, solutions of (2.2.2) are spatially homogeneous solutions of (2.2.1).

Linearizing (2.2.1) about $\mathbf{w} = \mathbf{w}_0$, we obtain the system

$$\begin{aligned} \mathbf{z}_t =& D\Delta\mathbf{z} + d\mathbf{f}(\mathbf{w}_0)\mathbf{z}, \quad \mathbf{x} \in \Omega,\ t > 0, \\ \nabla z_i.\mathbf{n} =& 0, \quad \mathbf{x} \in \partial\Omega,\ t > 0,\ i = 1, 2. \end{aligned} \tag{2.2.3}$$

Here $d\mathbf{f}(\mathbf{w}_0)$ is the (2×2) matrix defined by

$$(d\mathbf{f}(\mathbf{w}_0))_{i,j} = \frac{\partial f_i}{\partial w_j}(\mathbf{w}_0).$$

(To obtain (2.2.3), we set $\mathbf{w} = \mathbf{w}_0 + \mathbf{z}(\mathbf{x}, t)$, substitute into (2.2.1), and retain only linear terms in $\mathbf{z}$. The resulting linear system approximates the behaviour of (2.2.1) while $\mathbf{z} = \mathbf{w} - \mathbf{w}_0$ remains small.)

The trivial solution, $\mathbf{z} = 0$, is asymptotically stable if and only if every solution of (2.2.3) decays to zero as $t \to \infty$.

Let $\phi_j(\mathbf{x})$ denote the jth eigenfunction of $-\Delta$ on Ω with no-flux boundary conditito ns. That is,

$$\begin{aligned} \Delta\phi_j + \lambda_j\phi_j =& 0, \quad \mathbf{x} \in \Omega, \\ \mathbf{n}.\nabla\phi_j =& 0, \quad \mathbf{x} \in \partial\Omega, \end{aligned}$$

for scalars λ_j satisfying

$$0 = \lambda_0 < \lambda_1 < \lambda_2 \ldots.$$

The determination of the pairs (ϕ_j, λ_j) is a standard problem [22], [51]. The differential operator $-\Delta$, with no-flux boundary conditions, is self-adjoint in $L_2(\Omega)$, that is

$$\int_\Omega -\Delta\psi_1.\psi_2 \, dx = \int_\Omega -\Delta\psi_2.\psi_1 \, dx,$$

(where $dx = dx_1 dx_2 \dots dx_n$), and it is easy to see that, for each j,

$$\lambda_j = \frac{\int_\Omega |\nabla \phi_j|^2 \, dx}{\int_\Omega \phi_j^2 \, dx}$$

is nonnegative.

We may suppose without loss of generality that the ϕ_js are normalized so that

$$\|\phi_j\|_{L_2(\Omega)} = 1.$$

Moreover, the set ϕ_j form an orthogonal basis for $L_2(\Omega)$ and any function may be expanded as a Fourier series or eigenfunction expansion

$$u(\mathbf{x}) = \sum_{j=0}^{\infty} u_j \phi_j(\mathbf{x}).$$

Compare this with (1.4.13).

Using these preliminaries, we may solve (2.2.3) by expanding our solution $\mathbf{z}$ via

$$\mathbf{z} = \sum_{j=0}^{\infty} \mathbf{s}_j(t)\phi_j(\mathbf{x}) \tag{2.2.4}$$

where each $\mathbf{s}_j(t) \in \mathbf{R}^2$. This is simply **separation of variables** employed for (2.2.3). Substituting (2.2.4) into (2.2.3) and equating the coefficients of each ϕ_j, we have

$$\frac{d\mathbf{s}_j}{dt} = B_j \mathbf{s}_j$$

where B_j is the matrix

$$B_j = [d\mathbf{f}(\mathbf{w}_0) - \lambda_j D].$$

Now the trivial solution $\mathbf{z} = 0$ of (2.2.3) is asymptotically stable if and only if each $\mathbf{s}_j(t)$ decays to zero. This is equivalent to the condition that each B_j has two eigenvalues with negative real parts, that is that each B_j is a **stability matrix**.

On the other hand, if any B_j has an eigenvalue with positive real part, then $|\mathbf{s}_j|$ can grow exponentially and hence so will $\mathbf{z}$. Clearly, in this case, $\mathbf{z} = 0$ is unstable to arbitrary perturbations which are not orthogonal to ϕ_j.

It is worth pointing out here that (2.2.3) may be written as

$$\mathbf{z}_t + A\mathbf{z} = 0,$$

where A is the linear differential operator defined on a suitable subset of $L_2(\Omega)$, incorporating the boundary conditions, so that $-A\mathbf{z}$ agrees with the right-hand side of (2.2.3) (c.f. (1.4.12)). The spectrum of A consists of eigenvalues, μ_k ($k = 0, 1, 2, \ldots$), such that each $-\mu_k$ is an eigenvalue of a matrix B_j, for some $j = 0, 1, 2, \ldots$. Thus, by examining the behaviour of the eigenvalues of the B_js, we are simply determining the spectrum of the linear operator in (2.2.3).

If the parameters (d_1, d_2) are such that some B_j has an eigenvalue with zero real part, then as they are varied locally, the stability of $\mathbf{z} = 0$ will switch, and our preliminary investigations in section 1.4 suggest that this may reflect a bifurcation of some inhomogeneous equilibrium from the trivial solution $\mathbf{w} = \mathbf{w}_0$ for (2.2.1). Moreover, we expect that near any such bifurcation, the spatial structure of the nontrivial solutions will reflect that of ϕ_j, the critical eigenmode.

In Box G, below, we describe a general and rigorous approach to bifurcation theory which may be applied in situations such as this, assuring us that the change of stability is indeed accompanied by a bifurcation of inhomogeneous equilibrium. Alternatively, one may go for an asymptotic expansion, after the manner of section 1.4, and obtain the local bifurcation diagram near to the critical values for (d_1, d_2). For the moment, let us simply note that if B_j has an eigenvalue with positive real part, then small perturbations of $\mathbf{w} = \mathbf{w}_0$ in (2.2.1) will grow and develop spatially homogeneous pattern.

To proceed, we must be more explicit. Let

$$d\mathbf{f}(\mathbf{w}_0) = \begin{bmatrix} \alpha & \beta \\ \gamma & \delta \end{bmatrix}.$$

The eigenvalues, σ, of B_0 satisfy

$$\sigma^2 - \sigma(\alpha + \delta) + (\alpha\delta - \beta\gamma) = 0$$

and must lie in the left half of the complex plane, since $\mathbf{w}_0$ is a stable solution of (2.2.2), by hypothesis. Hence we have

$$\alpha + \delta < 0 \tag{2.2.5}$$

$$\alpha\delta - \beta\gamma > 0. \tag{2.2.6}$$

The eigenvalues, σ, of B_j satisfy

$$\sigma^2 - \sigma(\alpha + \delta - \lambda_j(d_1 + d_2)) + h(\lambda_j) = 0, \tag{2.2.7}$$

where h is given by

$$h(\lambda) = \lambda^2 d_1 d_2 - \lambda(d_1\delta + d_2\alpha) + (\alpha\delta - \beta\gamma). \tag{2.2.8}$$

Roots σ_1, σ_2 of (2.2.7) satisfy

$$\sigma_1 + \sigma_2 = \alpha + \delta - \lambda(d_1 + d_2) < 0,$$

since λ, d_1, d_2 are nonnegative and (2.2.6) holds. Hence, a root may only cross the imaginary axis at the origin, in which case, $h(\lambda_j) = 0$. This is a necessary condition for the change of stability, and hence for the associated bifurcation to occur. The exchange of stability is correspondingly reflected in the behaviour of perturbations in the jth eigenmode. The condition $h(\lambda_j) = 0$ may be visualized as a curve in the d_1, d_2 plane. Then, we can vary d_1 and d_2 and observe growth of small perturbations in the solution of (2.2.3) as we cross any such curve in the first quadrant of the d_1, d_2 plane.

From (2.2.5), we suppose without loss of generality that $\alpha < 0$. Then, if $d_1 = 0$, we always have $h > 0$, so all the B_j's are stability matrices. If $\delta \leq 0$ also, it is easy to see that h in (2.2.8) is the sum of one positive and two nonnegative terms. Thus, all the B_j's are always stability matrices and no bifurcation is possible. Hence, in order to proceed further, we assume

$$\alpha < 0, \quad \delta > 0.$$

Now, $h(\lambda_j) = 0$ if and only if

$$\lambda_j d_1 = \frac{\alpha\delta - \gamma\beta - \alpha\lambda_j d_2}{(\delta - \lambda_j d_2)}$$

which represents a curve in the $d_1 - d_2$ plane while $0 \leq d_2 \leq \frac{\delta}{\lambda_j}$. Figure 2.1 depicts such curves. Some simple algebra shows that each intersects the others exactly once. Moreover, since the B_j's are stable when $d_1 = 0$, we see that the trivial solution, $z = 0$, for (2.2.3) becomes unstable in the jth eigenmode as the point (d_1, d_2) moves downwards over the corresponding j^{th} curve in Figure 2.1.

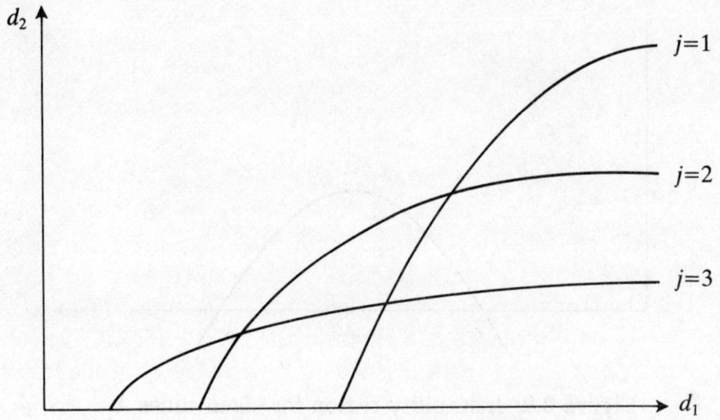

Figure 2.1: Neutral stability curves.

An alternative approach is to suppose d_1 and d_2 are fixed and allow Ω to vary. For domains with smooth boundaries, the eigenvalues λ_j vary continuously with Ω (see [51], for example). Hence, this current approach is equivalent to regarding the eigenvalues $\lambda = \lambda_j$ to be nonnegative variables and looking for conditions upon d_1 and d_2 which allow instability to be present whenever eigenvalues lie in some nonempty window amongst the positive reals. Specifically, we know that stability changes when $h(\lambda) = 0$. As $\lambda \to \infty$, we have $h > 0$ in (2.2.8), whilst (2.2.6) implies $h(0) > 0$ also.

The function h possesses two positive real roots, $\lambda\pm$, if and only if both

$$\begin{aligned} d_1\delta + d_2\alpha &> 0 \\ (d_1\delta + d_2\alpha)^2 &> 4d_1d_2(\alpha\delta - \beta\gamma). \end{aligned} \tag{2.2.9}$$

The roots $\lambda\pm$ are given by

$$\lambda\pm = \frac{[d_1\delta + d_2\alpha] \pm \sqrt{(d_1\delta + d_2\alpha)^2 - 4d_1d_2(\alpha\delta - \beta\gamma)}}{d_1d_2}.$$

When this is the case, B_j will have an eigenvalue, σ, in the right- hand side of the complex plane if and only if λ_j lies in (λ_-, λ_+). The real part of σ varies continuously with λ_j (see Figure 2.2), and the trivial solution, $\mathbf{w} = \mathbf{w}_0$, of (2.2.1) will be unstable in the jth eigenmode.

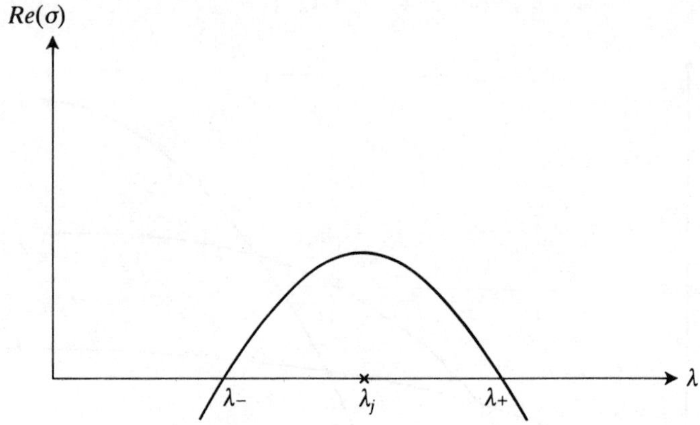

Figure 2.2: Instability region for eigenvalues λ_j.

If $d_1, d_2, \alpha, \beta, \gamma, \delta$ are varied, we may push the window around, or shrink it to a point and see it vanish as $\lambda\pm$ coalesce and go complex.

At any choice of parameters where the stability is neutral, a bifurcation analysis along the lines suggested in Box G or section 1.4 shows that non-trivial equilibria for (2.2.1) bifurcate from $\mathbf{w} = \mathbf{w}_0$. We illustrate this by considering the following example.

Example

To illustrate the action of Turing instability, we consider a model similar to those employed by Gierer and Meinhardt [49] in their considerations of pattern formation in morphogenesis.

In their work, over a number of years, Gierer and Meinhardt produced numerical solutions to such problems on a variety of domains [69]. The spatially structured equilibria display a variety of patterns, and all can be considered to have been born at a Turing instability, as certain parameters are varied.

We shall continue to let $\mathbf{w},\ \mathbf{w}_n, \ldots$ denote entities taking values in $\mathbf{R}^2$, with coordinates $(u, v)^T,\ (u_n, v_n)^T, \ldots$.

Consider (2.2.1) where

$$\mathbf{f} = \mathbf{f}(u, v) = \begin{bmatrix} p - uv^2 \\ q - v + uv^2 \end{bmatrix}. \tag{2.2.10}$$

Here the constants p and q are such that

$$p > q > 0, \text{ and } (p+q)^3 > (p-q).$$

Then, in the notation introduced earlier, we have

$$\mathbf{w}_0 = \begin{bmatrix} u_0 \\ v_0 \end{bmatrix} = \begin{bmatrix} p/(p+q)^2 \\ p+q \end{bmatrix},$$

$$d\mathbf{f}(\mathbf{w}_0) = \begin{bmatrix} \alpha & \beta \\ \gamma & \delta \end{bmatrix} = \begin{bmatrix} -(p+q)^2 & \frac{-2p}{p+q} \\ (p+q)^2 & \frac{p-q}{p+q} \end{bmatrix},$$

$$\det d\mathbf{f}(\mathbf{w}_0) = \alpha\delta - \beta\gamma = (p+q)^2 > 0,$$

$$\text{trace}\, d\mathbf{f}(\mathbf{w}_0) = \alpha + \delta = \frac{p - q - (p+q)^3}{p+q} < 0.$$

Clearly, $\alpha < 0$ and $\delta > 0$, so we expect to encounter diffusive instability when d_2 is small relative to d_1.

Let us choose the domain $\Omega = (0, \pi) \subset \mathbf{R}^n$, so that the eigenmodes of $-\Delta$, with no-flux boundary conditions, are simply

$$\phi_j = \cos jx, \quad \lambda_j = j^2, \quad j = 0, 1, 2, \ldots$$

(Here we have not normalized the ϕ_js in $L_2(\Omega)$.)

We also set

$$\begin{aligned} d_1 &= 1, \\ d_2 &= d, \end{aligned}$$

and seek a change of stability, for $\mathbf{w}_0$ (and the associated bifurcation) as d is decreased so that we move across one of the neutral stability curves depicted in Figure 2.1.

Now we note that $h = 0$ (in (2.2.8)) when

$$d = g(\lambda) = \frac{\lambda\delta - (p+q)^2}{\lambda(\lambda - \alpha)}, \tag{2.2.11}$$

which is positive for

$$\lambda > \frac{(p+q)^2}{\delta} = \frac{(p+q)^3}{(p-q)}.$$

Suppose (2.2.11) holds when $d = d^*$ and $\lambda = \lambda_j = j^2$, for some integer j, and that $g(\lambda_i) < d^*$, for all integers $i \neq j$. This simply says that as d decreases, from large positive values, the stability of the equilibrium, $\mathbf{w}_0$, is lost in the jth-mode first. As d decreases further, below d^*, there will be subsequent instabilities (and associated bifurcations), but it is the first one that is the most important. (The resultant bifurcating spatially inhomogeneous equilibria turn out to be stable as solutions of (2.2.1).)

Set

$$d = d^* - \varepsilon,$$

where ε is small, and expand any equilibrium solutions via

$$\mathbf{w}(x) = \mathbf{w}_0 + \sum_{n=1}^{\infty} \varepsilon^{\xi n} \mathbf{w}_n(x).$$

Here ξ is some positive constant still to be determined, controlling the respective scaling of d and $\mathbf{w}$; (c.f. the examples in section 1.4). Next we substitute for d and $\mathbf{w}$ into the steady-state system associated with (2.2.1).

The nonlinearity $\mathbf{f}$ in (2.2.10) may be expanded via its Taylor series, so that

$$\begin{aligned}\mathbf{f}(\mathbf{w}) = \mathbf{f}(\mathbf{w}_0) &+ \varepsilon^{\xi} d\mathbf{f}(\mathbf{w}_0).w_1 \\ &+ \varepsilon^{2\xi}(d\mathbf{f}(\mathbf{w}_0).\mathbf{w}_2 + \frac{1}{2}u_1^2\mathbf{f}_{uu} + u_1v_1\mathbf{f}_{uv} + \frac{1}{2}v_1^2\mathbf{f}_{vv}) \\ &+ \varepsilon^{3\xi}(d\mathbf{f}(\mathbf{w}_0).\mathbf{w}_3 + \frac{1}{2}u_1u_2\mathbf{f}_{uu} \\ &\quad + (u_1v_2 + u_2v_1)\mathbf{f}_{uv} + \frac{1}{2}v_1v_2\mathbf{f}_{vv} \\ &\quad + \frac{1}{6}u_1^3\mathbf{f}_{uuu} + \frac{1}{2}u_1^2v_1\mathbf{f}_{uuv} + \frac{1}{2}u_1v_1^2\mathbf{f}_{uvv} + \frac{1}{6}v_1^3\mathbf{f}_{vvv}) \\ &+ O(\varepsilon^{4\xi}).\end{aligned} \tag{2.2.12}$$

Here, $\mathbf{f}_{uu}$, $\mathbf{f}_{uv}, \ldots$ denote the vectors obtained by partially differentiating $\mathbf{f}$ component wise, evaluated at $\mathbf{w}_0$.

In fact

$$\mathbf{f}_{vv} = -2u_0\begin{bmatrix}1\\-1\end{bmatrix}, \ \mathbf{f}_{uv} = -2v_0\begin{bmatrix}1\\-1\end{bmatrix}, \ \mathbf{f}_{uvv} = -2\begin{bmatrix}1\\-1\end{bmatrix},$$

$$\mathbf{f}_{uu} = \mathbf{f}_{uuu} = \mathbf{f}_{uuv} = \mathbf{f}_{vvv} = 0,$$

so (2.2.12) is not as daunting as it first appears.

Now we equate powers of ε in our expansion of (2.2.1), and seek to determine $\mathbf{w}_i$ successively. To expedite matters, we set $\xi = 1/2$, since we suspect that the bifurcation is of pitchfork type, with a total of three equilibria when $\varepsilon > 0$ (and $\mathbf{w}_0$ is unstable). This indeed turns out to be the case, as described below; but the reader may wish to keep $\xi > 0$, unspecified, and continue in a manner analogous to that in the second example in section 1.4, and obtain $\xi = 1/2$ as a necessary choice for him or herself.

Let L denote the differential operator defined by

$$L\mathbf{w} = \left\{ \begin{bmatrix} 1 & 0 \\ 0 & d^* \end{bmatrix} \frac{\partial^2}{\partial x^2} + d\mathbf{f}(\mathbf{w}_0) \right\} \mathbf{w},$$

for suitable functions $\mathbf{w} : (0, \pi) \to \mathbf{R}^2$.

Then to $O(\varepsilon^{1/2})$ we have

$$L\mathbf{w}_1 = 0, \quad x \in (0, \pi); \quad \mathbf{w}_1 = 0, \quad x = 0, \pi.$$

Thus

$$\mathbf{w}_1 = A \cos jx.\mathbf{w}^*,$$

where $\mathbf{w}^*$ is the eigenvector in the null space of

$$B_j = \left\{ d\mathbf{f}(\mathbf{w}_0) - \lambda_j \begin{bmatrix} 1 & 0 \\ 0 & d^* \end{bmatrix} \right\}$$

given by

$$\mathbf{w}^* = \begin{bmatrix} u^* \\ v^* \end{bmatrix} = \begin{bmatrix} 2p/(p+q) \\ -j^2 - (p+q)^2 \end{bmatrix};$$

and A is a real constant still to be determined.

To $O(\varepsilon)$ we have

$$L\mathbf{w}_2 = (u_0 v_1^2 + 2v_0 u_1 v_1) \begin{bmatrix} 1 \\ -1 \end{bmatrix}, \quad x \in (0, \pi);$$

$$\mathbf{w}_2 = 0, \quad x = 0, \pi.$$

This has a solution, unique up to the addition of constant multiples of $\cos jx.\mathbf{w}^*$. We have

$$\begin{aligned} \mathbf{w}_2 = & \begin{bmatrix} u_2 \\ v_2 \end{bmatrix} \\ = & \frac{A^2(u_0 v^{*2} + 2v_0 u^* v^*)}{2(p+q)^2} \begin{bmatrix} 1 \\ 0 \end{bmatrix} \\ & + \frac{A^2(u_0 v^{*2} + 2v_0 u^* v^*)}{2((\alpha - 4j^2)(\delta - 4d^* j^2) - \beta\gamma)} \begin{bmatrix} -4j^2 d^* - 1 \\ 4j^2 \end{bmatrix} \cos 2jx \\ & + B\mathbf{w}^* \cos jx, \end{aligned}$$

for some constant B.

To $O(\varepsilon^{3/2})$ we have

$$\begin{aligned} L\mathbf{w}_3 =& v_{1\ xx}\begin{bmatrix}0\\1\end{bmatrix} - (u_0v_1v_2 + 2v_0(u_1v_2+u_2v_1) + u_1v_1^2)\begin{bmatrix}1\\-1\end{bmatrix}, \\ & x \in (0,\pi), \\ \mathbf{w}_3 =& 0, \quad x = 0, \pi. \end{aligned}$$

The solvability condition for this last equation demands that the right-hand side be orthogonal to $\mathbf{w}^* \cos jx$. Here, we mean orthogonal as fuctions in $L_2((0,\pi),\mathbf{R}^2)$. That is, if $\chi(x) = (\chi_1(x), \chi_2(x))^T$ and $\psi(x) = (\psi_1(x), \psi_2(x))^T$, then χ is orthogonal to ψ if and only if

$$\int_0^\pi \chi^T\psi\, dx = \int_0^\pi \chi_1(x)\psi_1(x) + \chi_2(x)\psi_2(x)\, dx = 0.$$

Thus, we have the condition

$$\begin{aligned} 0 =& v^* \int_0^\pi v_{1\ xx} \cos jx\, dx \\ & + (u^* - v^*)\int_0^\pi (u_0v_1v_2 + 2v_0(u_1v_2 + u_2v_1) + u_1v_1^2)\cos jx\, dx. \end{aligned}$$

The $B\mathbf{w}^* \cos jx$ term in $\mathbf{w}_2$ does not affect this last expression (once the integrals are calculated), and we are left with a cubic polynomial in the unknown A. We leave the evaluation of this expression to the reader. It has the form

$$0 = A(A_0^2 - A^2). \qquad (2.2.13)$$

Hence, as d decreases through the critical value, d^*, two nontrivial equilibria bifurcate from $\mathbf{w}_0$ according to

$$\mathbf{w} = \mathbf{w}_0 \pm (d^* - d)^{1/2}|A_0|\mathbf{w}^* \cos jx + O(d^* - d).$$

Having obtained the local bifurcation behaviour for this example, let us return to the more general question of the stability properties enjoyed by the nontrivial bifurcating solutions.

Although we may envisage linearizing the system about the bifurcating solution, it is not a straightforward matter to perform the spectral analysis for the resulting linear operator, especially if the system is a large one. However, we may note that the bifurcation is linked to a change of stability of the trivial solution in a subspace spanned by the appropriate (critical) eigenfunction. If the trivial solution remains stable to perturbations in all other eigenmodes, then we expect that such stability will be inherited by the bifurcating solutions. Thus, close to bifurcation, we need only check to see if the bifurcating solutions appear as stable solutions with respect to perturbations in the critical eigenmode.

Let us make the above notions more precise. We visualize the solutions of the reaction-diffusion system as trajectories, or orbits within an infinite-dimensional function space (each point corresponding to the instantaneous $\mathbf{x}$-dependence of $\mathbf{w}$; each orbit parameterized by time). An **invariant manifold** is a subset of the space which contains the whole orbit (both forwards and backwards in time) belonging to each and every point in it. Thus, if we have a solution which lies in an invariant manifold at time t_0, then it must do so for all $t \in \mathbf{R}$, by definition.

At the constant, trivial solution, $\mathbf{w}_0$, the perturbations within the critical eigenmode are represented by a linear subspace. The **centre manifold** is defined to be the invariant manifold, containing $\mathbf{w}_0$, that is tangential to the critical eigenmode there. Intuitively, one can see that the centre manifold contains all solutions arising from infinitesimal perturbations of $\mathbf{w}_0$ in the critical eigenmode. When bifurcation occurs, the exchange of stability of $\mathbf{w}_0$ and the appearance of nontrivial solutions take place in the centre manifold. Theoretically, and asymptotically in a neighbourhood of the bifurcation point, we may seek to analyse the dynamical behaviour on the centre manifold. If the trivial state, $\mathbf{w}_0$, remains stable in all other eigenmodes then, close to $\mathbf{w}_0$, arbitrary solutions will quickly drop down towards the centre manifold, and their long-time behaviour will mirror that of nearby solutions within the manifold itself. By obtaining a picture of the dynamics on the finite-dimensional centre manifold, we are effectively able to carry out both the bifurcation and stability analysis simultaneously.

Carr's book [7] provides an excellent introduction to this subject, and theoretical and practical applications to parabolic systems are given in [35].

To illustrate these ideas in action, we shall indicate two methods by which we may obtain the dynamic bifurcation behaviour. The first is close to the theoretical construction of the centre manifold, whilst the second is more pragmatic, and is an extension of the bifurcation analysis used in section 1.4 as well as earlier in the current section.

Let us consider the problem discussed in section 1.4:

$$u_t = u_{xx} + L^2 u(1-u), \quad x \in (0,1), \quad u(0) = u(1) = 0, \quad t > 0.$$

When $L = \pi$, we saw how a nonconstant, nonnegative solution bifurcated from the trivial zero solution. The critical eigenmode is spanned by $\sin \pi x$. Let us use an eigenfunction expansion to represent the centre manifold as follows. For u in the centre manifold, close to 0, we write

$$u = u_1 \sin \pi x + u_2(u_1) \sin 2\pi x + u_3(u_1) \sin 3\pi x + \cdots.$$

Here we use the coefficient u_1 to parameterise the centre manifold near zero. The functions $u_j(u_1)$ reflect the deviation of the centre manifold from the critical eigenmode. As $u \to 0$ in the centre manifold, we require

$$\lim_{u_1 \to 0} \frac{u_j(u_1)}{u_1} = 0 \text{ for } j = 2, 3, \ldots$$

so that u becomes tangent to the critical eigenmode. Now we allow u_1 to be a function of t, and substitute for u in the original problem. Projecting the resulting expression on to successive eigenmodes, we seek asymptotic expressions for the $u_j(u_1)$, valid for u_1 small. We obtain

$$\frac{du_1}{dt} \sin \pi x = (L^2 - \pi^2) u_1 \sin \pi x - K_1 u_1^2 \sin \pi x + O(u_1 u_2, \ldots)$$

$$\frac{du_2}{du_1} \frac{du_1}{dt} = (L^2 - 4\pi^2) u_2 \sin 2\pi x - K_2 u_1^2 \sin 2\pi x + O(u_1 u_2, \ldots)$$

$$\frac{du_3}{du_1} \frac{du_1}{dt} = (L^2 - 9\pi^2) u_3 \sin 3\pi x - K_3 u_1^2 \sin 3\pi x + O(u_1 u_2, \ldots).$$

Here

$$K_n = \frac{\int_0^1 \sin^2 \pi x \sin n\pi x \, dx}{\int_0^1 \sin^2 n\pi x \, dx}.$$

Substituting for du_1/dt from the first of these into the others, we can solve for u_2, $u_3, \ldots$ to order u_1^2 by seeking power series solutions. We obtain

$$u_2 = \frac{-K_2 u_1^2}{L^2 + 2\pi^2} + O(u_1^3),$$

$$u_3 = \frac{-K_3 u_1^2}{L^2 + 7\pi^2} + O(u_1^3) \ldots.$$

Substituting back into the du_1/dt equation, we have

$$\frac{du_1}{dt} = (L^2 - \pi^2) u_1 - K_1 u_1^2 + O(u_1^3).$$

Hence, since $K_1 > 0$, for $L > \pi$, there is a stable positive equilibrium for u_1. For $L < \pi$, the trivial solution, $u_1 = 0$, is stable. Hence the nonnegative, nontrivial bifurcating solution depicted in Figure 1.4 is locally stable.

In this example, there was not much feed-back from the higher eigenmodes to the critical one. Things are a little less trivial if we attempt the same programme for the pitchfork-type bifurcations exhibited by our example earlier in this section.

We leave this as an exercise for those who feel up to some prolonged algebra. For such problems, our second approach is more straightforward.

Let us reconsider the above example. In section 1.4, we analysed the bifurcation of equilibria by introducing the small parameter, ε, via $L = \pi + \varepsilon$. Then we expanded u as a series in powers of ε^α, for some suitable positive power α. In this example, the choice $\alpha = 1$ was the appropriate one. We have the asymptotic series

$$u = \varepsilon u_1 + \varepsilon^2 u_2 + \varepsilon^3 u_3 + \cdots.$$

In the present setting, we generalize the previous method by allowing the u_j to depend on a scaled time variable as well as x. More precisely, we set

$$\tau = |\varepsilon^\beta| t,$$

where β is some positive constant, so that τ represents a **long time scale** (t is $O(|\varepsilon|^{-\beta})$ when τ is $O(1)$).

Now we let

$$u_j = u_j(x, \tau), \quad j = 1, 2, \ldots$$

and seek an equation for u_1, which possesses τ-dependent solutions which are nonconstant, yet uniformly bounded. The power balancing in ε will determine β.

Substituting into the equation, we have

$$\begin{aligned} 0 = &- |\varepsilon|^\beta \varepsilon \frac{du_1}{d\tau} + \varepsilon(u_{1xx} + \pi^2 u_1) \\ &+ \varepsilon^2(u_{2\,xx} + \pi^2 u_2 + 2\pi u_1 - \pi^2 u_1^2) \\ &+ O(\varepsilon^3, \varepsilon^2 |\varepsilon|^\beta). \end{aligned}$$

If $\beta = 0$, then, to $O(\varepsilon)$, we have a linear equation for u_1 which cannot yield any bounded nonconstant solutions. Hence $\beta > 0$, and imposing the boundary conditions, $u_1 = 0$ at $x = 0, 1$, we have

$$u_1 = A(\tau) \sin \pi x,$$

for some function $A(\tau)$. If $\beta > 1$, then the ε^2 terms, together with the boundary conditions ($u_2 = 0$ at $x = 0, 1$), yield

$$A^2 \pi K_1 - 2A = 0,$$

as in section 1.4 (here we have applied the solvability condition). K_1 is as defined earlier in our first approach to this problem.

Clearly, this gives constant values for A, corresponding to the bifurcating equilibria. Thus, to obtain nonconstant solutions for A we choose $\beta = 1$, so that the solvability condition yields

$$A^2 \pi K_1 - 2A = \operatorname{sgn}(\varepsilon)\frac{dA}{d\tau}.$$

Hence, for $\varepsilon > 0$ we have time-dependent solutions leaving a neighbourhood of $A = 0$, and tending to the stable rest point $A = 2/\pi K_1$ as $\tau \to \infty$. For $\varepsilon < 0$ the origin is stable.

Hence, we obtain the same result found earlier via the centre manifold method, and, by rescaling the respective u_1 variables as well as time, one may check that the two methods have both produced the same time-dependent approximation in the critical eigenmode.

One may apply this method easily to our pattern formation example discussed earlier (given by (2.2.1) and (2.2.10)). There, the bifurcation equation (2.2.13) came from the application of the solvability condition to the $O(\varepsilon^{3/2})$ terms. Since we expanded $\mathbf{w}$ in powers of $\varepsilon^{1/2}$, we must choose

$$\tau = |\varepsilon| t$$

so that the $du_1/d\tau$ term is also $0(\varepsilon^{3/2})$. Then we obtain a τ-dependent differential equation in place of (2.2.13). One may check that the resulting bifurcating solutions are stable. This is left as an exercise (just add in the extra term $A_\tau \cos jx.\mathbf{w}^*$ to the right-hand side of the $\mathbf{w}_3$ equation).

Box G: Bifurcation theory

We pause here to describe a more rigorous approach to bifurcation theory, as applied to partial differential equations. There are now many excellent texts on this subject ([8], [23], for example) and its applications (for example, [70] Chapter 13).

In the following, we shall outline the major steps involved in the **Liapunov-Schmidt** approach to bifurcation. Here, the action takes place in Banach spaces, and the theory rests on ideas from functional analysis. The pay-off comes in having a general theory which guarantees bifurcation under certain generic hypotheses. The result is a scalar equation of the form $g(\alpha, \lambda) = 0$, where solutions $\alpha \in \mathbf{R}$ are in one-to-one correspondence with solutions, u, of the original bifurcation problem (λ is, as usual, the bifurcation parameter).

For example,

$$g(\alpha, \lambda) = \alpha(\lambda - \alpha^2)$$

represents a standard **pitchfork** bifurcation, having three solutions when $\lambda > 0$, and only the trivial solution for $\lambda \geq 0$.

However, the function g is only obtained implicitly near the bifurcation point $(\alpha, \lambda) = (0,0)$. Thus, some means of analysing the local qualitative behaviour of g must be sought. In texts such as [23], a standard procedure is presented which enables the calculation of the successive derivatives of g so that its local behaviour is known. In Chapter 1, section 1.4, we achieved this goal by introducing a small parameter, ε, and expanding both λ and the solution, u, in suitable power series. Clearly, this *ad hoc* approach relies on more subtle processes present within the analysis. In particular, it represents only an approximation to the rigorous results provided by the Liapunov-Schmidt technique.

In a given bifurcation problem, one may choose from a variety of approaches. The particular technique employed should take into account the degree of information required of the solution, as well as one's own personal preferences.

We shall now outline the usual Liapunov-Schmidt reduction of a bifurcation problem. There are many generalizations [23], but we shall content ourselves with a simple one-parameter problem.

Let X and Y denote Banach spaces, and

$$\Phi : X \times \mathbf{R} \rightarrow Y$$

be a smooth mapping such that $\Phi(0,0) = 0$. Suppose we wish to solve

$$\Phi(u, \lambda) = 0 \tag{1}$$

for $u \in X$ as a function of λ near $(0,0)$. Let L be the differential, or linearization, of Φ with respect to u at the origin:

$$Lu = \lim_{h \rightarrow 0} \frac{\Phi(hu, 0) - \Phi(0,0)}{h}.$$

In the applications that we shall have in mind, (1) is a nonlinear elliptic problem (the steady-state problem for some reaction-diffusion system, for instance). Then X and Y are subspaces of $L_2(\Omega)$, for some domain Ω in $\mathbf{R}^n$. $L_2(\Omega)$ is a Hilbert space (a Banach space with norm generated by an inner product). Its inner product is given by

$$< u, v > = \int_\Omega uv^* \, d\mathbf{x},$$

where v^* denotes the complex conjugate of v. (Although we are interested in real valued functions, it is conventional to perform spectral analysis in the complexification of the underlying spaces: for example, real matrices may have complex eigenvectors.)

Next we must make some assumptions regarding L.

An operator $A : X \to Y$ is **Fredholm** if the nullspace, $\ker A$, is a finite dimensional subspace of X, and the range, range A, is a closed subspace of Y having finite co-dimension. For Fredholm operators, there exist closed subspaces M and N of X and Y, respectively, such that

$$\begin{aligned} X &= \ker A \oplus M \\ Y &= N \oplus \text{range } A. \end{aligned} \tag{2}$$

This means that for each $u \in X$, there exists unique elements $v \in \ker A$, and $w \in M$ such that $u = v + w$ (and similarly, a unique decomposition of Y into N and range A). By definition, we have

$$\dim N = \text{codim range } A.$$

The **Fredholm index** of A is given by

$$\dim \ker A - \dim N.$$

We assume henceforth that the operator L, introduced above, is a Fredholm operator of index zero.

In general, the decomposition of Y in (2) requires a little care. However, in the current context, L is an elliptic differential operator, and it can be shown that (2) is valid [22], [51]. The point is that for such operators, we have

$$(\text{range } L)^{\perp} = \ker L^*,$$

where L^* denotes the adjoint operator of L. This is defined by $L^*x^* = y^*$ whenever x^* satisfies $< Lx, x^* > = < x, y^* >$ for all $x \in X$, for some suitable y^* (note that the domain of L^* is generally quite distinct from that of L). Also, we have used $(\text{range } L)^{\perp}$ to denote the orthogonal complement of range L, defined by

$$\{y \in Y : < y, v > \text{ for all } v \in \text{range } L\}.$$

The characterization of $(\text{range } L)^{\perp}$ is known as the **Fredholm alternative**, and provides a rigorous justification for the kind of solvability conditions introduced in Box C, in Chapter 1.

In the present setting, it allows us to make the choice

$$N = \ker L^*,$$

and, since we require L to be of index zero, we need only check that

$$\dim\ker L = \dim\ker L^*.$$

This is trivial if L happens to be self-adjoint.

We shall assume that $\dim\ker L = 1$, since this is generally the case; however, the Liapunov-Schmidt theory still holds good when $\dim\ker L > 1$, and we shall discuss this briefly later.

Now we introduce the projection $E : Y \to \text{range } L$ associated with the decomposition in (2), and rewrite (1) as a pair of equations

$$E\Phi(u, \lambda) = 0 \tag{3}$$

$$(I - E)\Phi(u, \lambda) = 0. \tag{4}$$

We use the decomposition $u = v + w$ where $v \in \ker L$ and $w \in M$ and solve (3) for w as a function of v and λ. We have

$$E\Phi(v + w; \lambda) = 0. \tag{5}$$

The combined map

$$F(v, w, \lambda) = E\Phi(v + w, \lambda)$$

takes $\ker L \times M \times \mathbf{R}$ into range L. Moreover, the differential with respect to w of F at the origin is $EL = L$. However, when the domain of L is restricted to M, we have $L : M \to \text{range } L$, which is invertible. Now we can apply the implicit function theorem to conclude that (5) can be solved for w near $(v, \lambda) = (0, 0)$. That is,

$$w = W(v, \lambda),$$

which depends smoothly on v and λ.

Next we substitute for w in (4) to obtain

$$(I - E)\Phi(v + w(v, \lambda), \lambda) = 0. \tag{6}$$

This last equation is, in effect, the bifurcation equation, valid for (v, λ) in the domain of w. We have reduced the original problem (1), in an infinite-dimensional space, to solving (6) for $v \in \ker L$, which is of finite dimension.

We may rewrite (6) as a real-valued bifurcation equation as follows: introduce $v_0 \in \ker L$ and $v_0^* \in \ker L^*$ such that

$$\|v_0\|_{L_2} = \|v_0^*\|_{L_2} = 1.$$

(Since $\dim\ker L = \dim\ker L^* = 1$, $\ker L$ and $\ker L^*$ are precisely the linear, one-dimensional, subspaces spanned by v_0 and v_0^* respectively.)

Now we set $v = \alpha v_0$ for $\alpha \in \mathbf{R}$, and define

$$g(\alpha, \lambda) = < v_0^*, \Phi(\alpha v_0 + w(\alpha v_0, \lambda), \lambda) >,$$

so that (6) implies

$$0 = g(\alpha, \lambda)$$

which represents our bifurcation equation defined on a neighbourhood of the origin in (α, λ)-space, (i.e. $\mathbf{R}^2$).

If the linear operator, L, had been Fredholm, with index zero, and

$$\dim \ker L = r,$$

then (6) could be reduced to r algebraic equations in r unknowns (representing some suitable coordinates for $\ker L$).

If L were Fredholm, with nonzero index, then (6) would represent either an under-determined or over-determined system of equations.

In Golubitsky and Schaeffer [23], the authors analyse the behaviour of g by deriving expressions for its derivatives. This, coupled with an argument allowing for the truncation of Taylor series, enables them to discuss the qualitative structure of the local bifurcation.

Having described the ideas behind Liapunov-Schmidt theory (providing a rigorous basis for bifurcation analysis), we refer the interested reader to [23], for a more detailed exposition (and, in particular, for an introduction to the use of symmetry in simplifying the calculations).

2.3 Transition layers

For some systems, the methods of the previous section are inappropriate. It is often the case that spatially inhomogeneous equilibria are far from any uniform equilibria, and bifurcation analysis is of no use. In this case, we must find some alternative techniques.

Generally, we are faced with a coupled system of m nonlinear (steady-state) elliptic equations, and even for one-dimensional domains, their solution would require the construction of orbits in a $2m$-dimensional phase space.

Often however, one or more of the diffusion coefficients are small compared to others. When this happens, we can seek solutions which ignore the corresponding diffusion terms everywhere except for a discrete number of points. In neighbourhoods of these special points, the state variables are allowed to change sharply, while elsewhere, we can get an approximate solution by solving a system of reduced order.

We illustrate these ideas with an example drawn from population biology. Here, the state variables u and v denote nonnegative population densities of species of prey and predator respectively. The diffusion terms reflect the ability of individuals to disperse randomly within the domain (see section 1.2 and Box A). The reaction terms are population supply terms due to births and deaths.

Consider the steady-state problem

$$\varepsilon^2 u_{xx} + u(h(u) - v) = 0 \tag{2.3.1}$$
$$v_{xx} + v(u - g(v)) = 0 \tag{2.3.2}$$

$$\text{for } x \in (0, L),$$
$$u_x = v_x = 0 \text{ at } x = 0, L,$$

where g and h are of the qualitative forms depicted in Figure 2.3.

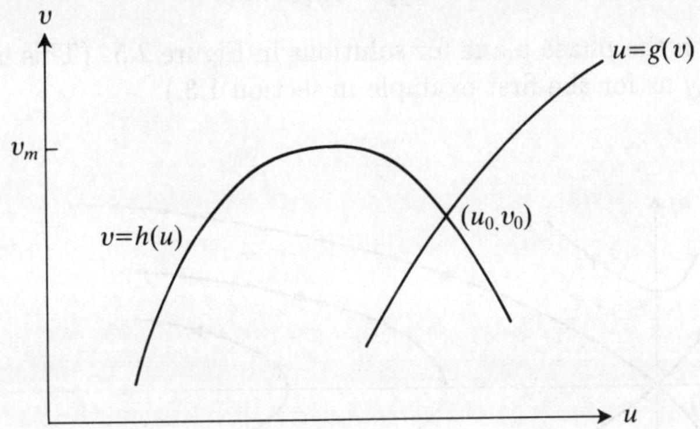

Figure 2.3: Nullclines for (2.3.1-2).

For example, we could choose $h(u) = u(1-u)$ and $g(v) = 0.5 + v$.

Next, suppose that ε tends to zero. In the limit, we may solve (2.3.1) algebraically, so that either

$$u = 0$$
$$u = k_-(v)$$
$$u = k_+(v)$$

where k_- and k_+ are the pseudo inverses of the function h defined on $[0, v_m]$; $k_+(v)$ represents the upper branch, while $k_-(v)$ gives the lower one; see Figure 2.4.

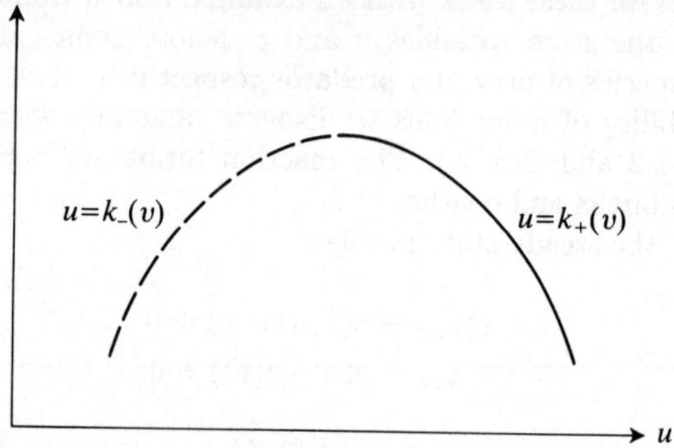

Figure 2.4: The functions k_+ and k_-.

Now we can look for solutions of (2.3.2).
If $u \equiv 0$, we have

$$v_{xx} - vg(v) = 0. \tag{2.3.3}$$

We sketch the phase plane for solutions in Figure 2.5. (This is done in the same way as for the first example in section 1.3.)

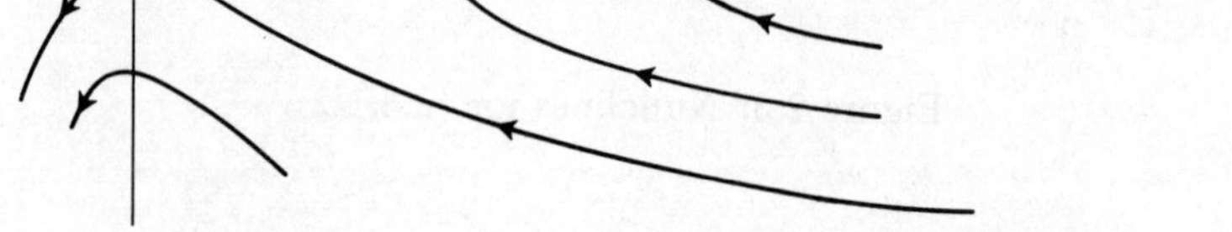

Figure 2.5: The phase plane for (2.3.3).

Clearly, there can be no nontrivial, nonnegative solutions satisfying

$$v_x = 0 \text{ at } x = 0, L,$$

for any fixed value of $L > 0$.

Similarly, we look for solutions of (2.3.1-2) on the upper branch $u = k_+(v)$. We have

$$v_{xx} + v(k_+(v) - g(v)) = 0. \tag{2.3.4}$$

Figure 2.6 depicts the behaviour of the nonlinearity and Figure 2.7 shows the phase plane.

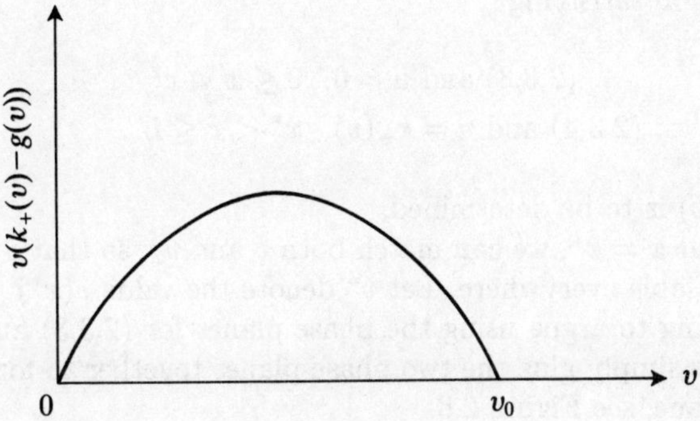

Figure 2.6: The nonlinearity in (2.3.4).

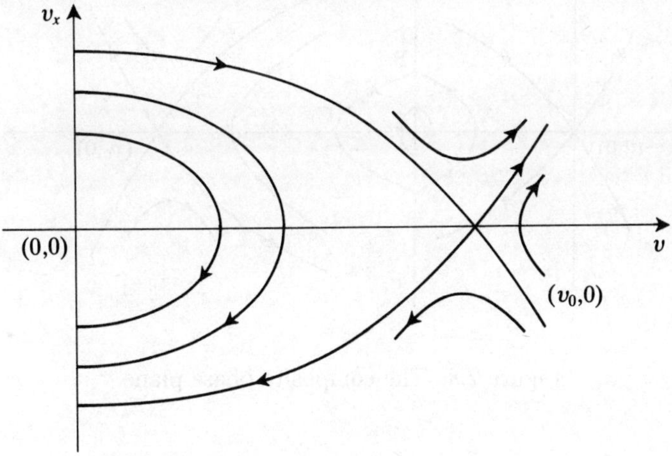

Figure 2.7: The phase plane for (2.3.4).

Again, there are no nonnegative, nontrivial solutions satisfying the boundary conditions.

(We have not considered the case $u = k_-(v)$, since this branch does not give rise to stable solutions; but one may easily check that when $u = k_-(v)$, there are no nontrivial solutions satisfying the boundary conditions.)

Obviously, we must be more ingenious if we are to construct solutions of (2.3.1-2), even in the limit $\varepsilon = 0$!

One possibility that presents itself is to allow the solution to *jump* between solving the equations (2.3.3) and (2.3.4). More precisely, we shall look for a solution satisfying

$$\begin{aligned} (2.3.3) \text{ and } u = 0, &\quad 0 \le x < x^* \\ (2.3.4) \text{ and } u = k_+(v), &\quad x^* < x \le L. \end{aligned} \tag{2.3.5}$$

Here, $x^* \in (0, L)$ is to be determined.

Moreover, at $x = x^*$, we can match both v and v_x, so that v is continuously differentiable everywhere. Let v^* denote the value $v(x^*)$.

It is revealing to argue using the phase planes for (2.3.3) and (2.3.4). For v^* fixed, we simply glue the two phase planes together to form a composite phase plane; see Figure 2.8.

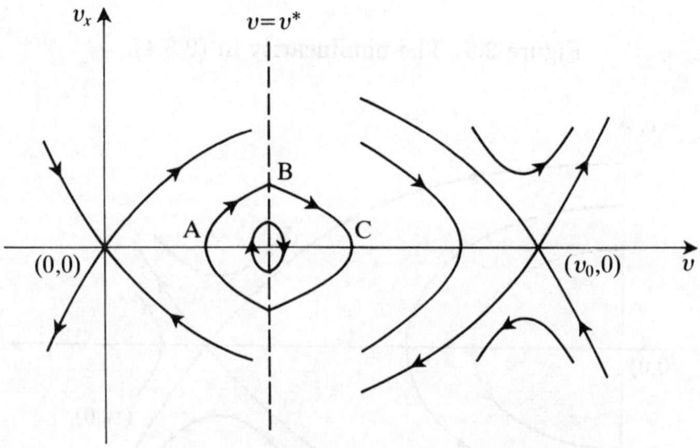

Figure 2.8: The composite phase plane.

Now we look for all orbits of the form ABC in Figure 2.8, which are candidates for our solution. Here, we must have $x = 0$ when the orbit segment starts out at A (where $v_x = 0$); and $x = L$ when the orbit reaches C (where again $v_x = 0$). Assuming for the moment that this is the case, x^* is fixed by the condition $v(x^*) = v^*$ when the orbit passes through B.

To be more precise, let $v^* \in (0, v_0)$ be fixed (recall that (u_0, v_0) is the rest point for (2.3.1-2) – see figure 2.3). Define

$$f(v) = \begin{cases} -vg(v) & v < v*, \\ -v(g(v) - k_+(v)) & v > v^*, \end{cases}$$

and

$$F(v) = \int_0^v f(w)\, dw.$$

Along orbits in the composite phase plane, we have

$$0 = v_{xx} + f(v).$$

Multiplying by v_x and integrating from 0 to x, we obtain

$$\frac{v_x^2(x)}{2} + F(v(x)) - F(v(0)) = 0.$$

Thus, since $v_x \geq 0$ along the orbit ABC, we have

$$x = \frac{1}{\sqrt{2}} \int_{v(0)}^{v(x)} \frac{dv}{(F(v(0)) - F(v))^{1/2}}.$$

In particular, we have the so-called time-map (see section 1.4),

$$L = \frac{1}{\sqrt{2}} \int_{v(0)}^{v(L)} \frac{dv}{(F(v(0)) - F(v))^{1/2}}. \tag{2.3.6}$$

Here, $v(L)$, the value of v at $x = L$ (at C in Figure 2.8), satisfies $v^* < v(L) < v_0$, and

$$F(v(L)) = F(v(0)). \tag{2.3.7}$$

Two possibilities present themselves, depending upon the precise form of the phase plane:

(i) If $F(1) > 0$, then the unstable manifold of (0,0), in the first quadrant, never reaches $v = v_0$. Instead, it passes through the v-axis, at $v = v_1$, and returns to the origin as a stable manifold, in the fourth quadrant; see Figure 2.9(a). Such an orbit is called **homoclinic**. For every $v(0) \in (0, v^*)$, there exists $v(L) \in (v^*, v_1)$, satisfying (2.3.7), and an orbit of the form ABC in Figure 2.8. Then (2.3.6) gives the corresponding domain length L.

(ii) If $F(1) < 0$, then the unstable manifold of (0,0), in the first quadrant, remains there at least as far as the line $v = v_0$. The saddle point at $(v_0, 0)$ possesses a homoclinic orbit intersecting the v-axis at

$v = v_2 \in (0, v^*)$; see Figure 2.9(b). For each $v(0) \in (v_2, v^*)$ there exists $v(L) \in (v^*, v_0)$, satisfying (2.3.7), and, again, L is given by (2.3.6).

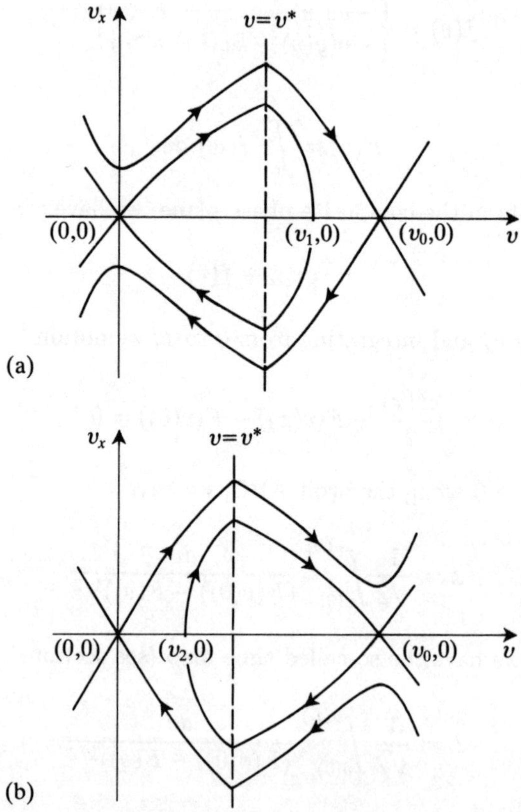

Figure 2.9: Two possible configurations for the composite phase plane: (a) $F(1) > 0$, (b) $F(1) < 0$.

In either case, as $v(0)$ varies below v^*, we can regard $v(L)$ and L as functionals of $v(0)$, satisfying (2.3.6) and (2.3.7). In the original statement of our problem, L is assumed fixed, so ideally, we would like to invert (2.3.6), and write $v(0)$ (and hence $v(L)$) as functionals of L. Unfortunately, such an expression is not available. However, it can be shown that the time-map, (2.3.6), is a monotonically decreasing function of $v(0)$ (using (2.3.5) to write $v(L)$ implicitly in terms of $v(0)$), [72]. The separate cases (i) and (ii) above, correpond to Figures 2.10(a) and 2.10(b). If necessary, we may evaluate (2.3.6) numerically, but the time-map analysis of [72] can be employed to provide a rigorous justification for the behaviour depicted in Figure 2.10.

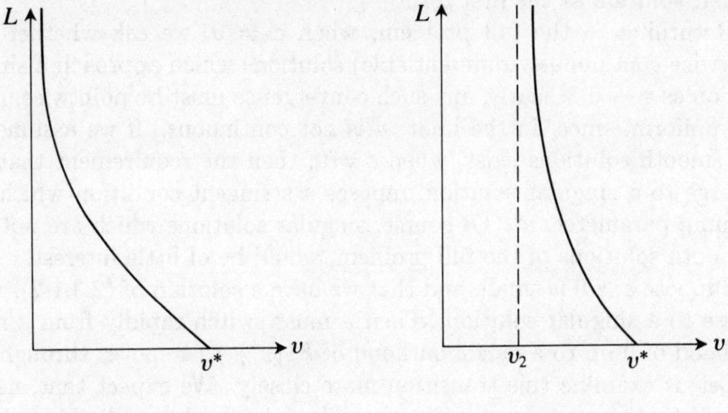

(a) $F(1) > 0$ (b) $F(1) < 0$

Figure 2.10: The time-map given by (2.3.6).

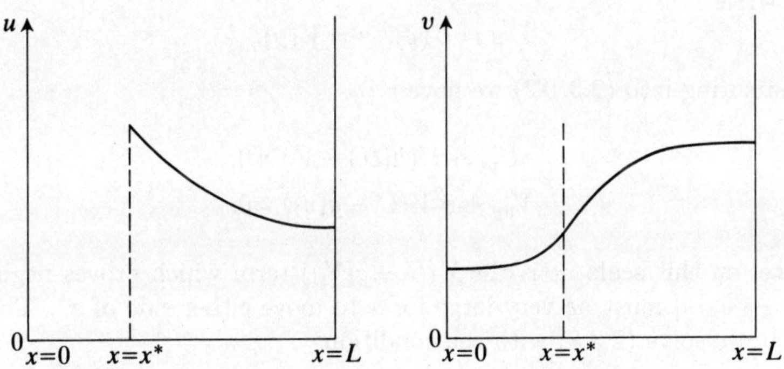

Figure 2.11: The singular solution for (2.3.1-2), when $\varepsilon = 0$.

So far, we have shown that when $\varepsilon = 0$, and v^* is suitably fixed, (2.3.1-2) possesses a solution of the form (2.3.5). Here x^* is fixed by the particular solution itself, and u has a simple jump discontinuity there (see Figure 2.11). We refer to such a solution as **singular**, since it possesses such a jump. When $\varepsilon \neq 0$, the singular solution does not have enough smoothness for the left-hand side of (2.3.1) to be meaningful at $x = x^*$, so we cannot seek solutions via a regular expansion in powers of ε (using the singular solution as the first term).

Returning to the full problem, when $\varepsilon \neq 0$, we ask whether there are (twice continuously differentiable) solutions which approach a singular solution as $\varepsilon \to 0$. Clearly, any such convergence must be pointwise, rather than uniform, since, in the limit, u is not continuous. If we assume that such smooth solutions exist, when $\varepsilon \neq 0$, then the requirement that they converge to a singular solution imposes a stringent condition which fixes the jump-parameter, v^*. Of course, singular solutions which are not close to smooth solutions of the full problem, would be of little interest.

Suppose $\varepsilon > 0$ is small, and that we have a solution of (2.3.1-2), which is close to a singular solution. Then u must switch rapidly from a neighbourhood of 0 up to a neighbourhood of $k_+(v^*)$, as x moves through x^*.

Let us examine this transition more closely. We expect that, near to $x = x^*$, the $\varepsilon^2 u_{xx}$ term, in (2.3.1), will no longer be negligible. On the other hand, v and v_x are continuous, even for the singular solution, so we ought to have $v \approx v^*$ for our smooth solution, near to $x = x^*$.

Set

$$y = \frac{x - x^*}{\varepsilon},$$

and write

$$u = U(y), \quad v = V(y).$$

Substituting into (2.3.1-2) we have

$$\begin{aligned} U_{yy} + U(h(U) - V) &= 0, \\ V_{yy} + \varepsilon^2 V(U - g(v)) &= 0. \end{aligned} \tag{2.3.8}$$

Hence, on this scale, it is the $V(U - g(V))$ term which proves negligible. As $\varepsilon \to 0$, $|y|$ must be very large for x to move either side of x^*. Thus, as $\varepsilon \to 0$, we solve (2.3.8) with the conditions

$$\begin{aligned} (U, V) &\to (0, v^*) \text{ as } y \to -\infty, \\ (U, V) &\to (k_+(v^*), v^*) \text{ as } y \to +\infty, \end{aligned}$$

so that in the limit the **inner** solution, (U, V), is matched with the **outer** solution, (u, v), which satisfies (2.3.5). When this is the case, our smooth solution is approximated by the (outer) singular solution away from $x = x^*$. On the x-scale the switch in u, approximated by the inner solution, takes place in a small neighbourhood of x^*. (More precisely, most of the action in the solution of (2.3.8) will take place while y is $O(1)$, as $\varepsilon \to 0$; which means that $x - x^*$ is $O(\varepsilon)$.) Such a neighbourhood is called an **internal transition layer**.

As $\varepsilon \to 0$, we can solve (2.3.8) asymptotically, by setting $\varepsilon = 0$. We have

$$V \equiv v^*,$$

and

$$U_{yy} + U(h(U) - v^*) = 0,$$

$$U \to 0 \text{ as } y \to -\infty,$$
$$U \to k_+(v^*) \text{ as } y \to +\infty.$$

Multiplying by U_y and integrating over $\mathbf{R}$, with respect to y, we obtain

$$0 = G(v^*)$$

where

$$G(v^*) = \int_{\mathbf{R}} U_y U(h(U) - v^*)\, dy = \int_0^{k_+(v^*)} U(h(U) - v^*)\, dU.$$

This determines v^* uniquely. For h as depicted in Figure 2.3, we have $G(0) > 0$, while $G(v_m) < 0$, and by direct calculation,

$$\frac{dG}{dv^*} = -k_+(v^*) < 0.$$

Hence, we have a unique $v^* \in (0, v_m)$.

So far, we have shown that the condition fixing v^* is a necessary one for (2.3.1-2) to have a solution close to the singular one. In fact, it is also sufficient. The argument, [16],[50], is a technical one which employs a version of the implicit function theorem. We shall not go into this here, preferring to stress the useful nature of the transition layer analysis. The singular solutions (and their transition profiles) may be compared directly to numerical solutions of (2.3.1-2). In more complicated systems, a cursory transition layer approach is often useful in preparing numerical methods and confirming numerical or experimental data.

We shall meet transition layers again in Chapter 3, where similar arguments are used to construct travelling waves. This again is taken a stage further in Chapter 4, where we meet transition layers in $\mathbf{R}^3$ and on two-dimensional surfaces.

The present analysis, for reaction-diffusion equations in one space dimension, dates back to the ideas of Paul Fife in the mid 1970's (see [16],[17], for example). There are many points of interest in such problems which we have not stressed here, and we refer the interested reader to [11] and the references therein.

From one point of view, the transition layer approach may be thought of as the construction of the first-order terms in a matched asymptotic expansion of the solution (u, v).

Matched asymptotic expansions have been used for some time in applied mathematics (see [52]), and we discuss a few applications of this technique in Box H.

Box H: Matched asymptotic expansions

When a problem contains a small parameter, say ε, we may try to analyse the behaviour of the solution in the limit $\varepsilon \to 0$. In doing so, our aim is to obtain an approximation to the true solution, valid for small ε. Often the solution, say u, may be expanded as a regular series in some power of ε. That is, an expansion of the form

$$u = u_0(x) + \varepsilon^{\alpha} u_1(x) + \varepsilon^{2\alpha} u_2(x) + \cdots, \quad x \in \Omega, \tag{1}$$

(like those we used in our bifurcation analysis in sections 1.5 and 2.2). Here $\alpha > 0$ is some constant.

We substitute such an expression into the original problem and seek to solve for the $u_k(x)$ by equating terms in successive powers of ε. However, sometimes such a series solution may exhibit **singular** behaviour. To be more precise, (1) is **regular** at x provided the quotients

$$\frac{u_{i+1}(x)}{u_i(x)}$$

are bounded. When this is not the case, the

$$\varepsilon^{(i+1)\alpha} u_{i+1}(x)$$

term is no longer negligible, compared to the $\varepsilon^{i\alpha} u_i(x)$ term, and the process of equating terms of similar order, by which we solved for the u_i, is no longer justified.

Commonly, for differential equations, it is the case that simply setting $\varepsilon = 0$ (by which we obtain the first-order problem which must be solved for u_0) reduces the order of the problem. In such situations, it is not generally possible to force u_0 to satisfy all of the boundary conditions. We then suspect that our expansion will no longer be valid near to the boundaries where any mismatch occurs. For example, consider the problem

$$\varepsilon u_{xx} - 2u_x + \exp u = 0, \; x \in (0, 1), \;\; u(0) = u(1) = 0. \tag{2}$$

Substituting (1) into (2), we see that u_0 must satisfy

$$2u_{0\,x} = \exp u_0, \quad x \in (0,1), \tag{3}$$

and, since (1) must satisfy the boundary conditions (for $\varepsilon = 0$), we also have

$$u_i(0) = u_i(1) = 0, \; i = 0, 1, 2 \ldots.$$

But the solution of (3) cannot satisfy both conditions. We have either $u_0(0) = 0$, in which case

$$u_0(x) = \ln\left(\frac{2}{2-x}\right), \tag{4}$$

or $u_0(1) = 0$, in which case

$$u_0(x) = \ln\left(\frac{2}{3-x}\right). \tag{5}$$

Whichever we choose, we expect that our regular expansion cannot approximate the solution close to both end points. In fact the true solution for this problem possesses a **boundary layer**, that is a region close to one of the end-points in which u changes rapidly so as to satisfy the appropriate boundary condition. Such a region will shrink to a single point, as $\varepsilon \to 0$, so it is perhaps not surprising that our regular expansion was not able to pick it up.

In the general situation, when a regular series becomes singular at a point, say x_0, we can try to obtain a new expansion which is valid in a local neighbourhood of x_0. To do so, we first rescale the independent variables (and possibly the dependent variables). The idea here is to stretch out the neighbourhood of x_0 by introducing a new independent variable of the form $y = (x - x_0)/\varepsilon^a$, for some $a > 0$. A good guide to choosing a is to make sure that the new first-order problem (obtained by setting $\varepsilon = 0$) contains at least one term not included in the old first-order problem, yet also has at least one term in common with it. This really means that we are now allowing a previously neglected term to dominate in the determination of the new expansion (and, hopefully, supply the behaviour that our old expansion lacked). The old expansion, valid away from x_0, is called the **outer** expansion, while the new expansion, in the stretched variable y, is called the **inner** expansion.

Having obtained both inner and outer expansions, they will usually contain some free constants which must be determined by **matching** the two expressions close to x_0. In reality, although our expansions are formally valid in separate regions, we need to assert that there is some region of

overlap, near x_0, on a scale between that of x and y. When such a region exists (and its existence is by no means trivial), we can determine any free constants by forcing agreement between our expansions on this intermediate scale. This is usually straightforward for the first-order terms in both series, but requires more care when a number of successive terms are to be matched. Let us return to our boundary layer problem which was left unresolved above.

Let us choose (4) as the first term in our outer regular expansion. Since $u_0(1) = \ln 2$, we look for an inner expansion valid near the end point $x = 1$. Set $y = (1-x)/\varepsilon^a$, and write $u = U(y)$ in (2). We have

$$\varepsilon^{1-2a}U_{yy} + 2\varepsilon^{-a}U_y + \exp U = 0.$$

If $a > 1$, then when $\varepsilon \to 0$, the U_{yy} term is the one of lowest order. If $a < 1$, then likewise, we are left with the U_y term only. Hence, the choice $a = 1$ is the natural one. Setting

$$U(y) = U_0(y) + O(\varepsilon)$$

we have

$$U_{0\ yy} + 2U_{0\ y} = 0,$$

which has the general solution

$$U_0(y) = A\exp(-2y) + B,$$

for constants A and B.

Now since our original problem demands $u(1) = 0$, we must impose the corresponding condition, $U(0) = 0$, upon U_0. Hence, we have

$$A + B = 0,$$

but no way to determine both constants completely. Now comes the matching. As $x \to 1$, $u_0(x) \to \ln 2$ (see (4)). For x close to 1, the corresponding value of the stretched variable, y, tends to ∞ as $\varepsilon \to 0$. Thus, for $u_0(x)$ and $U_0(y)$ to agree, in the limit $\varepsilon \to 0$, we impose

$$\lim_{y\to\infty} U_0(y) = \lim_{x\to 1} u_0(x) = \ln 2.$$

Hence, $B = \ln 2$, so that the behaviour of the solution close to $x = 1$ is given by

$$\ln 2\left(1 - \exp\left(\frac{-2(1-x)}{\varepsilon}\right)\right) + O(\varepsilon).$$

Often we can write our matched pair of expansions as a single composite expansion. In this example, we observe that each first-order term takes the value ln2 in the region where the other is valid. Hence, by adding them together, and subtracting an amount ln2, we obtain a uniformly valid asymptotic expansion;

$$u(x) = \ln\left(\frac{2}{2-x}\right) - \ln 2 \exp\left(\frac{2(x-1)}{\varepsilon}\right) + O(\varepsilon).$$

Suppose now that, instead of (4), we had chosen (5) as our outer expansion, satisfying the boundary condition at $x = 1$. Then, proceeding as before, we would have sought an inner expansion, valid near to $x = 0$, say, $u = U(y) = U_0(y) + O(\varepsilon)$, where $y = x/\varepsilon$. We would have had

$$U_{0\ yy} - 2U_{0\ y} = 0, \quad U(0) = 0,$$

and hence

$$U_0(y) = A \exp 2y$$

for some constant A, to be determined by matching $U_0(y)$ with (5).
However, for $x > 0$, the corresponding value of y tends to ∞ as ε tends to zero, so we would have required $u(0) = \ln 2/3$, in (5), to match the limit of $U_0(y)$ as $y \to \infty$. This last limit is undefined for $A \neq 0$, so no such match would have been possible.
In effect, the method would be telling us that (5) is the wrong choice for the outer expansion, and that we should try to place the boundary layer at $x = 1$ (as indeed we did earlier).
It is always reassuring to find analytical methods which are able to throw out any erroneous choices quickly!

2.4 Oscillatory patterns

Here, we shall consider a recently proposed mechano-chemical model for morphogenesis, [53],[54]. The aim is not only to introduce a more elaborate pattern-forming mechanism, but also to illustrate a different kind of bifurcating solution: namely, a solution exhibiting a spatial structure and temporal oscillations.
When oscillating solutions branch from stationary equilibria, we say that a **Hopf bifurcation** has occurred. This is familiar within the study of ordinary differential equations, where a limit cycle bifurcates from a rest

point. The rest point suffers a change of stability, as in the case of the bifurcation phenomenon discussed previously. However, in Hopf bifurcation, the spectrum of the linearized problem does not meet zero at the critical bifurcation point, but rather contains values which cross the imaginary axis with nonzero imaginary part. A simple example should suffice to illustrate these basic ideas.

Consider

$$\begin{aligned} x_t =& \lambda x + y - \alpha x(x^2 + y^2) \\ y_t =& - x + \lambda y - \alpha y(x^2 + y^2). \end{aligned} \tag{2.4.1}$$

Here, $\lambda \in \mathbf{R}$ is our bifurcation parameter, and $\alpha \in \mathbf{R}$ is fixed.

Linearising (2.4.1) about $(x, y) = (0, 0)$, we have

$$\begin{bmatrix} x_t \\ y_t \end{bmatrix} + A \begin{bmatrix} x \\ y \end{bmatrix} = 0$$

where

$$A = \begin{bmatrix} -\lambda & -1 \\ 1 & -\lambda \end{bmatrix}.$$

The eigenvalues of A are complex and are given by

$$-\lambda \pm i.$$

Thus, the origin switches from being asymptotically stable to being unstable as λ increases through zero. However, in (2.4.1), no new rest points appear when $\lambda > 0$.

In polar coordinates $x = r\cos\theta, y = r\sin\theta$, (2.4.1) becomes

$$\begin{aligned} r_t =& \lambda r - \alpha r^3 \\ \theta_t =& - 1. \end{aligned}$$

Thus, when $\text{sgn}(\lambda) = \text{sgn}(\alpha)$, there is a limit cycle (given by $r^2 = \lambda/\alpha$). If $\alpha > 0$, then the limit cycle is stable ($r_t > 0$ for $r \in (0, \lambda/\alpha)$, $r_t < 0$ for $r > \lambda/\alpha$) while, if $\alpha < 0$, then the limit cycle is unstable and exists only for negative λ. Figure 2.12 illustrates the results schematically.

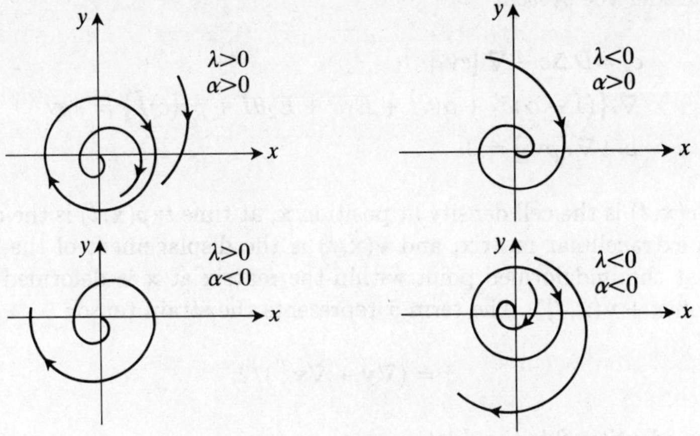

Figure 2.12: Solutions of (2.4.1).

When $\alpha > 0$, a stable limit cycle is born as λ increases through zero. This phenomenon is usually called **supercritical** Hopf bifurcation. When $\alpha < 0$, an unstable limit cycle collapses into the rest point as λ increases through zero. This is termed **subcritical** Hopf bifurcation.

There are many excellent texts on Hopf bifurcation, [7],[25]; its application in infinite dimensional evolutionary systems (e.e. nonlinear parabolic partial differential equations!) being of particular interest. The concepts in Box H and centre manifold theory, introduced in section 2.3, are essential to a full understanding of generalized Hopf bifurcation.

In the general case, the aim is to use centre manifold theory to reduce the problem to a finite dimensional one (canonically two dimensional) on the centre manifold. Then one can determine the super/subcriticality of the bifurcation by examining a system of ordinary differential equations resembling (2.4.1). These calculations are often alarmingly complex even for simple examples.

For our purposes, we shall just note that when the linearized operator possesses pure imaginary eigenvalues, we should anticipate the appearance of oscillatory, time dependent solutions.

Now we turn to a model for pattern formation in morphogenesis. After a brief statement of the problem, we shall reduce it to a simplified case where Hopf bifurcations are present.

In the following model, we consider a population of mobile embryonic cells, distributed on a viscoelastic extracellular matrix, which itself resists the deformations caused by the tractive forces exerted by the cells, [53],[54].

We consider the system

$$\begin{aligned} c_t =& D\Delta c - \nabla.[c\mathbf{v}_t] \\ \nabla.\{(1-\alpha)\tilde{\sigma}_t &+ \alpha\theta_t\tilde{I} + E_1\tilde{\sigma} + E_2\theta\tilde{I} + \rho\tau(c)\tilde{I}\} = s\rho\mathbf{v} \\ \rho_t+&\nabla.[\rho\mathbf{v}_t] = 0. \end{aligned}$$

Here, $c(\mathbf{x},t)$ is the cell density at position $\mathbf{x}$, at time t; $\rho(\mathbf{x},t)$ is the density of the extracellular matrix; and $\mathbf{v}(\mathbf{x},t)$ is the displacement of the matrix (so that the undeformed point within the matrix at $\mathbf{x}$ is deformed to the location $\mathbf{x}+\mathbf{v}(\mathbf{x},t)$). The term $\tilde{\sigma}$ represents the strain tensor

$$\tilde{\sigma} = (\nabla\mathbf{v} + \nabla\mathbf{v}^T)/2;$$

$\tilde{I}$ is the identity; θ is the dilation

$$\theta = \nabla.\mathbf{v};$$

while $\tau(c)$ represents the traction exerted by the cells upon the matrix. For simplicity, we shall assume τ is linear

$$\tau(c) = \tau_0 c,$$

although the more general case would not cause much extra work.

The first and third equations represent conservation of cell mass and matrix mass, given the advection due to matrix displacements. The second equation is of the form

$$\nabla.(\text{stress tensor}) + (\text{body force}) = 0,$$

with the terms $\tilde{\sigma}_t, \theta_t$ representing the contribution due to viscosity and $E_1\tilde{\sigma} + E_2\theta\tilde{I}$ to the elastic nature of the matrix.

Finally, $\alpha \in (0,1), E_1, E_2, D$, and s are positive constants.

There has to date been some work on its pattern-forming potential, although mainly in one dimensional problems. The simplification introduced below is made with an eye to problems in two- and three-dimensional domains.

We assume that the undisturbed equilibrium $(c,\rho,\mathbf{v}) \equiv (c_0,\rho_0,0)$ (for constants c_0,ρ_0), and the constants α, E_1, E_2, s are such that, nearby, the behaviour of the solution is dominated by c and $\mathbf{v}$, so that ρ may be taken to be constant (in fact, $\equiv 1$). By rescaling the problem, one can find sets of parameters where this is a more or less valid approximation, and this is left as an exercise to the interested reader.

Hence, we shall consider

$$c_t = D\Delta c - \nabla[c\mathbf{v}_t] \tag{2.4.2}$$
$$\nabla.\{(1-\alpha)\tilde{\sigma}_t + \alpha\theta_t\tilde{I} + E_1\tilde{\sigma} + E_2\theta\tilde{I} + \tau(c)\tilde{I}\} = s\mathbf{v}, \tag{2.4.3}$$

where $\mathbf{x} \in \Omega \subseteq \mathbf{R}^n$ and $t > 0$.
If $\Omega \neq \mathbf{R}^n$, suitable boundary conditions must be supplied:

$$\mathbf{v} = \nabla c.\mathbf{n} = 0, \quad x \in \partial\Omega, \quad t > 0,$$

where $\mathbf{n}$, as usual, denotes the outer normal to $\partial\Omega$. The system (2.4.2-3) is still rather complicated in more than one dimension, owing to the viscous and elastic terms present in the stress tensor in (2.4.3).
However, we can make things more tractable as follows:
Suppose we seek solutions on some bounded domain $\Omega \subseteq \mathbf{R}^3$. We define the scalar field ϕ via

$$\Delta\phi = \nabla.\mathbf{v}, \quad x \in \Omega,$$
$$\nabla\phi.\mathbf{n} = 0, \quad x \in \partial\Omega.$$

For a smooth $\mathbf{v}$, vanishing on $\partial\Omega$, ϕ is determined up to the addition of any constant, and may be written

$$\phi = \int_\Omega G(\mathbf{x},\mathbf{y})\nabla.\mathbf{v}(\mathbf{y},t)\,d\mathbf{y}$$

where G is the Green function for the Laplacian on Ω with no-flux boundary conditions (note that the solvability condition

$$\int_\Omega \nabla.\mathbf{v}\,dx = 0$$

is trivially satisfied).
Now we define

$$\mathbf{p} = \mathbf{v} - \nabla\phi$$

so that $\nabla.\mathbf{p} = 0$ and the sum

$$\mathbf{v} = \nabla\phi + \mathbf{p} \tag{2.4.4}$$

represents the unique decomposition of $\mathbf{v}$ into an irrotational part, $\nabla\phi$, and divergence free, rotational part, $\mathbf{p}$. It is easy to check that

$$\nabla.\tilde{\sigma} = \Delta(\nabla\phi) + \frac{1}{2}\Delta\mathbf{p}$$
$$\theta = \Delta\phi,$$

so (2.4.3) may be rewritten as

$$0 = \nabla\{\Delta\phi_t + (E_1 + E_2)\Delta\phi + \tau(c) - s\phi\} + \frac{1}{2}\Delta\{(1-\alpha)\mathbf{p}_t + E_1\mathbf{p}\}. \quad (2.4.5)$$

Now (2.4.5) represents the unique decomposition of the trivial vector field, 0, into irrotational and divergence free parts. Thus, each are identically zero, so that we have

$$\Delta\phi_t + (E_1 + E_2)\Delta\phi = s\phi - \tau(c) + k(t) \quad (2.4.6)$$

$$\Delta\{(1-\alpha)\mathbf{p}_t + E_1\mathbf{p}\} = 0, \quad \nabla.\mathbf{p} = 0. \quad (2.4.7)$$

We have integrated the irrotational part in order to obtain (2.4.6). The term arising from the integration, $k(t)$, is determined by demanding that (2.4.6) possesses a solution, ϕ. Note that if ϕ is shifted by an arbitrary constant, then $\mathbf{v} = \nabla\phi$ is not altered, so without loss of generality, we may seek a *normalized* solution satisfying

$$\int_\Omega \phi \, dx = 0.$$

Then, integrating (2.4.6) over Ω, we obtain

$$k(t) = \frac{-\int_\Omega \tau(c)\, dx}{\int_\Omega dx}.$$

In fact, this last condition is also sufficient to ensure that (2.4.6) possesses a (normalized) solution, ϕ (see Box C and Box G for solvability conditions and the Fredholm alternative).
Although the equation for $\mathbf{p}$ is decoupled from that for ϕ, at $\partial\Omega$ we have $\mathbf{p} = \mathbf{v} - \nabla\phi$; so that, if $\mathbf{v} = 0$ on $\partial\Omega$, then $\mathbf{p}$ is coupled to $\nabla\phi$ by the boundary values that must be imposed upon (2.4.7).
However, if we relax our original specification of the problem so that we only require

$$\mathbf{v}.\mathbf{n} = 0, \quad x \in \partial\Omega$$

(instead of $\mathbf{v} = 0$), we may seek a solution representing irrotational displacements by putting

$$\mathbf{v} \equiv \nabla\phi \quad \text{and} \quad \mathbf{p} \equiv 0.$$

In this case, we have the system

$$-\Delta[\phi_t + (E_1 + E_2)\phi] + s\phi = \tau(c) + k(t) \quad (2.4.8)$$

$$x \in \Omega, \ t > 0,$$

$$c_t = D\Delta c - \nabla[c\phi_t], \quad (2.4.9)$$

$$\nabla c.\mathbf{n} = \nabla\phi.\mathbf{n} = 0 \quad \text{for} \quad x \in \partial\Omega, \ t > 0. \quad (2.4.10)$$

Here again, we may assume without loss of generality that

$$\int_\Omega \phi \, dx \equiv 0,$$

so that

$$k(t) = \frac{-\int_\Omega \tau(c)\, dx}{\int_\Omega dx}.$$

Alternatively, we may expect that the solution of (2.4.7) will stay close to the equilibrium, $\mathbf{p} = 0$, except in a boundary layer, close to $\partial\Omega$, where $\mathbf{p}$ must match $-\nabla\phi$. In this case, (2.4.8-2.4.10) must represent a good approximation in the interior of Ω, to the full system (2.4.2,4,6,7) together with the boundary conditions $\mathbf{v} = \nabla c.\mathbf{n} = 0$. In consequence of these considerations, we shall consider the simplified system (2.4.8-10) for the remainder of this section.
Homogeneous steady states for (2.4.8-10) are given by

$$\begin{aligned} c =& N, \\ \phi =& 0, \end{aligned}$$

for any constant N. Here, $N > 0$ represents the average cell mass; notice that, since the right-hand side of (2.4.8) is in divergence form, the total cell mass is conserved for all time.
In order to seek nonconstant solutions, we must perform a stability analysis for the equilibria.
Fix $N > 0$ and set

$$c = N + \tilde{c}(\mathbf{x}, t)$$

where $\tilde{c}$ is assumed to be small, and

$$\int_\Omega \tilde{c}(\mathbf{x}, t)\, d\mathbf{x} = 0.$$

Substituting for c in (2.4.8-10), and retaining only linear terms in $\tilde{c}$ and ϕ, we obtain

$$\begin{aligned} \tilde{c}_t =& D\Delta\tilde{c} - N\Delta\phi_t \\ \Delta\phi_t =& -E\Delta\phi + s\phi - \tau'(N)\tilde{c} \end{aligned} \qquad x \in \Omega,\ t > 0, \qquad (2.4.11)$$

together with

$$\int_\Omega \tilde{c}\, d\mathbf{x} = \int_\Omega \phi\, d\mathbf{x} = 0, \qquad t > 0, \qquad (2.4.12)$$

$$\mathbf{n}.\nabla\tilde{c} = \mathbf{n}.\nabla\phi = 0, \qquad x \in \partial\Omega,\ \ t > 0. \qquad (2.4.13)$$

We have written $E = E_1 + E_2$ and $\tau'(N) = d\tau/dc$ at $c = N$.
Solutions for (2.4.11) satisfying the boundary conditions may be obtained from the eigenfunctions ψ_i of the problem

$$\begin{aligned}-\Delta\psi_i =& \lambda_i\psi_i, \qquad x \in \Omega,\\ \mathbf{n}.\nabla\psi_i =& 0, \qquad x \in \Omega.\end{aligned}$$

As usual (see section 2.2) we have $\psi_0 \equiv 1$ and $\lambda_0 = 0$, while $\lambda_i > 0$ and $\int_\Omega \psi_i \, d\mathbf{x} = 0$ for $i = 1, 2 \ldots$.
We set

$$\begin{bmatrix}\tilde{c}\\ \phi\end{bmatrix} = \psi_i(x)\begin{bmatrix}\overline{c}(t)\\ \overline{\phi}(t)\end{bmatrix} \tag{2.4.14}$$

for some integer $i \geq 1$, so that (2.4.12) and (2.4.13) are immediately satisfied.
Substituting into (2.4.11), we obtain

$$\begin{bmatrix}\overline{c}_t\\ \overline{\phi}_t\end{bmatrix} = \begin{bmatrix}-D\lambda_i + \tau'(N)N & -(E\lambda_i + s)N\\ \tau'(N)/\lambda_i & -(E\lambda_i + s)/\lambda_i\end{bmatrix}\begin{bmatrix}\overline{c}\\ \overline{\phi}\end{bmatrix}. \tag{2.4.15}$$

The determinant of the matrix in (2.4.15) is given by

$$\det = D(E\lambda_i + s) > 0, \quad \text{for } i = 1, 2, \ldots$$

whilst the trace may be positive or negative depending upon the parameter values. If the trace is negative, then the eigenvalues of the matrix have negative real part, and the solutions of (2.4.15) tend to zero as $t \to \infty$. If the trace is positive, then the eigenvalues of the matrix have positive real part, and the solutions of (2.4.15) grow as $t \to \infty$. Thus, if the trace is positive, the system (2.4.11)-(2.4.13) is unstable in the ith eigenmode (spanned by functions of the form (2.4.14)).
At criticality, the trace is zero, so

$$N\tau'(N) = \lambda_i D + E + s/\lambda_i \tag{2.4.16}$$

where $\lambda = \lambda_i$, $(i = 1, 2, \ldots)$. For such parameter values, the eigenvalues of the matrix in (2.4.15) are pure imaginary complex conjugates and we may expect that a Hopf bifurcation may occur.
Note that the right-hand side of (2.4.16), considered as a function of λ, has a minimum of $E + 2\sqrt{sD}$ at $\lambda = \sqrt{s/D}$ (see Figure 2.13). Thus, if $\tau'(N)N$ is greater than $E + 2\sqrt{sD}$, there is a *window* of possible values, λ, for which the steady state is unstable. If Ω is chosen appropriately, then one or more of the λ_i will be inside the window and oscillatory perturbations may be observed.

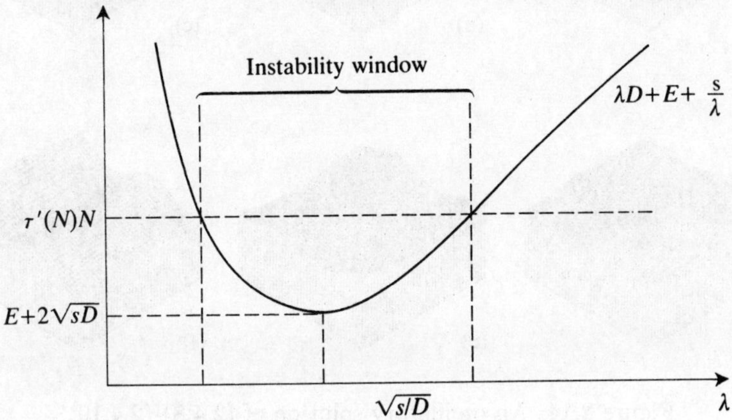

Figure 2.13: The instability window for (2.4.16).

Choosing Ω to be a square domain and suitable parameter values, we may solve (2.4.8)-(2.4.10) numerically. The resulting solutions are shown in Figures 2.14 and 2.15. Note that the time series of surfaces show growing oscillatory perturbations in time (as expected) as well as the spatial structure inherited from the corresponding $\psi_i(x)$.

The Hopf bifurcation, obtained as say $N\tau'(N)$ increases from zero until the first λ_i enters the instability window, may be sub or super critical; a full analysis employing centre manifold theory is beyond our present scope. Nevertheless, we have succeeded in demonstrating that spatio-temporal heterogeneous patterns may develop in a class of models originally derived for such a purpose.

Patterns in two and three dimensions are of ongoing current interest for such systems, and it is hoped that the analysis above will be developed further.

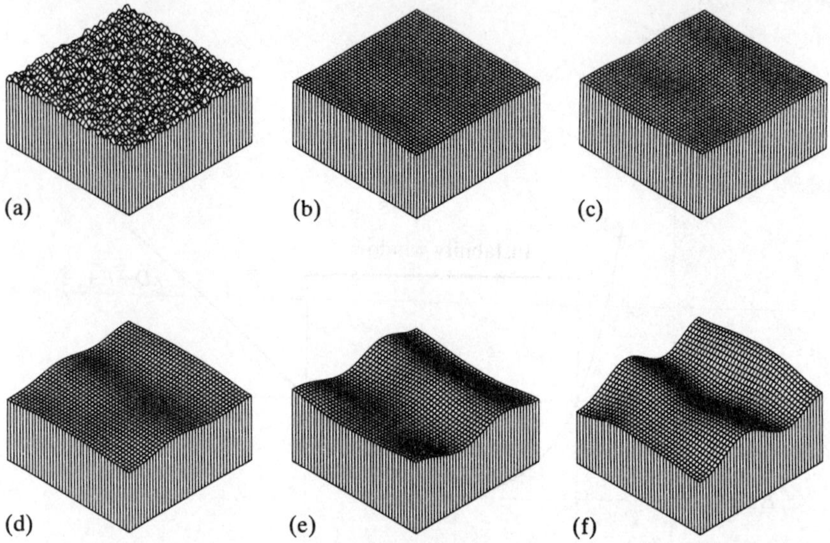

Figure 2.14: An oscillatory solution of (2.4.8)-(2.4.10).

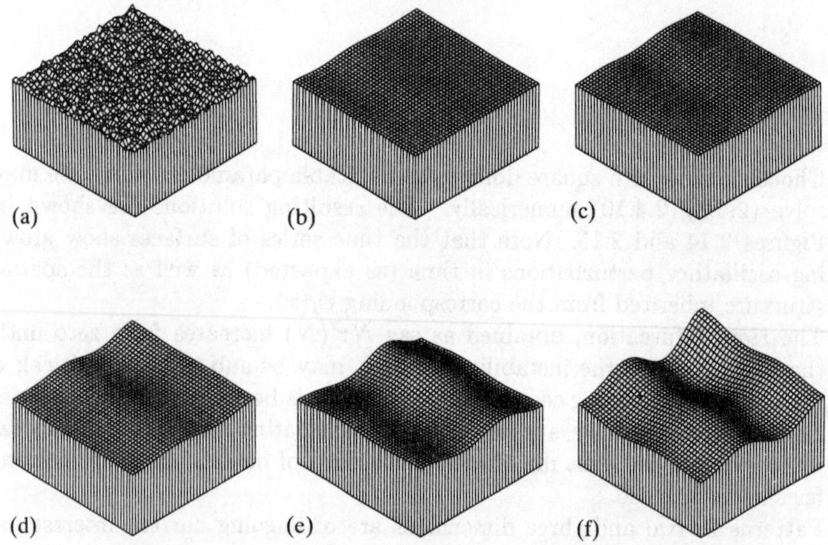

Figure 2.15: An oscillatory solution of (2.4.8)-(2.4.10).

Exercise 2.1
Find the bifurcation structure in a neighbourhood of the bifurcation points $u = 0, \lambda = \pi^2$ for the following problems:
a)

$$u_{xx} + \lambda \sin u = 0 \qquad x \in (0,1)$$

$$u(0) = u(1) = 0$$

b)

$$u_{xx} + \lambda(u + u^2) = 0 \qquad x \in (0,1)$$

$$u(0) = u(1) = 0$$

c)

$$u_{xx} + \lambda(u + xu^2) = 0 \qquad x \in (0,1)$$

$$u(0) = u(1) = 0.$$

Exercise 2.2
Consider the problem

$$u_{xx} + \lambda \sin u = \delta \qquad x \in (0,1)$$

$$u(0) = u(1) = 0.$$

Here the parameters λ and δ are both real, with $0 \leq |\delta| << 1$. Set $\lambda = \pi^2 + \epsilon$ and analyse the bifurcation from the trivial solution $u = 0, \epsilon = 0, \delta = 0$ by writing

$$u = A(\delta, \epsilon) \sin \pi X + W(x)$$

where $w \to 0$ faster than A as $\delta \to 0, \epsilon \to 0$ independently sketch the result in the (ϵ, A) plane, for $-1 << \delta < 0$, $\delta = 0$, $0 < \delta << 1$.

Exercise 2.3
In section 2.2, immediately prior to Box G, we analysed the bifurcation and stability of equilibria, generalizing the previous asymptotic expansion method of section 1.4 by allowing the coefficients to depend on a scaled time variable as well as a spatial variable. More precisely, we set

$$\tau = |\varepsilon^{\beta}|t,$$

where β was some positive constant, so that τ represented a long time scale. Seeking an equation for the first-order nontrivial solution, which possessed τ-dependent solutions which were nonconstant, yet uniformly bounded, the power balancing in ε determined β.

Apply this method to the system (2.2.1) and (2.2.10). The bifurcation equation (2.2.13) came from the application of the solvability condition to the $O(\varepsilon^{3/2})$ terms. Since we expanded $\mathbf{w}$ in powers of $\varepsilon^{1/2}$, choose

$$\tau = |\varepsilon| t$$

so that the $du_1/d\tau$ term is also $O(\varepsilon^{3/2})$. Obtain a τ-dependent differential equation for the coefficient $A(\tau)$ of $\cos jx.\mathbf{w}^*$.
Show that the resulting bifurcating solutions are stable.

Exercise 2.4
Consider the system

$$\frac{dx}{dt} = xy + \alpha x^3$$
$$\frac{dy}{dt} = -y + \beta x^2 + \gamma x^2 y.$$

What can be said about the stability of the rest point (0,0)? Show that the linearization has a stable manifold along the y-axis, and a degenerate manifold along the x-axis.
For the full system seek a centre manifold, tangent to the x-axis at (0,0), on which the stability is to be determined, by setting $y = h(x)$ locally along it, where $h(x) = O(x^2)$ for x small. Show that

$$\frac{dh}{dx}(x)\{xh(x) + \alpha x^3\} = -h(x) + \beta x^2 + \gamma x^2 y,$$

and

$$\frac{dx}{dt} = xh(x) + \alpha x^3$$

on the centre manifold.
Deduce that

$$h(x) = \beta x^2 + O(x^4)$$

and hence that the origin is asymptotically stable (unstable) if $\alpha + \beta < 0$ (> 0).
If $\alpha + \beta = 0$, show that the origin is asymptotically stable (unstable) if $\gamma\beta < 0$ (> 0).

Exercise 2.5
A model for enzyme kinetics is given by

$$\frac{dy}{dt} = -y + (y + c)z$$
$$\varepsilon\frac{dz}{dt} = y - z(y + 1).$$

Here ε, small, and $0 < c < 1$ are constants.
For $\varepsilon = 0$ we have $z = y/(1+y)$ and

$$\frac{dy}{dt} = \frac{-y(1-c)}{1+y}.$$

Use centre manifold theory to show that solutions of the full system remain close to solutions of this reduced equation, when ε is small, and $y(0)$ and $z(0)$ are small, as follows.
Introduce the fast time scale $\tau = t/\varepsilon$ and treat ε as a dependent variable (which happens to be constant). We have

$$\begin{aligned}\frac{dy}{d\tau} &= \varepsilon((y+c)(y-w)-y)\\ \frac{dw}{d\tau} &= y^2 - w - wy + \varepsilon((y+c)(y-w)-y)\\ \frac{d\varepsilon}{d\tau} &= 0,\end{aligned}$$

where $w \equiv y - z$.
Linearizing at (0,0,0), show that there is a centre manifold, tangent to the (y,ε) plane at (0,0,0), on which the behaviour is controlled by the nonlinear terms.
Show that

$$w = h(y,\varepsilon) = y^2 - (1-c)\varepsilon y + O(y^3, \varepsilon^3)$$

on this manifold.
Derive an equation for the dynamics when $w = h(y,\varepsilon)$.

Exercise 2.6
Consider a general initial value problem for the system

$$\begin{aligned}\frac{dy}{dt} &= -y + (y+c)z\\ \varepsilon\frac{dz}{dt} &= y - z(y+1)\end{aligned}$$

considered in the previous question. Using both t and the early (fast) time scale $\tau = t/\varepsilon$ to obtain a matched asymptotic expansion representing the solutions to $O(\varepsilon)$.
Can you obtain a higher-order expression?

Exercise 2.7
Consider the following system, often called the Brusselator,

$$\begin{aligned}u_t &= Du_{xx} + u^2v - (b+1)u + a\\ v_t &= dv_{xx} - u^2v + bu, \qquad 0 < x < L,\ t > 0,\end{aligned}$$

where a, b, D, and d are positive constants. We impose the boundary conditions

$$u(0,t) = u(L,t) = a$$
$$v(0,t) = v(L,t) = b/a.$$

Clearly $u \equiv a,\ v \equiv b/a$ is an equilibrium.
a) Show that this equilibrium is stable if and only if $b < 1+a^2$ for the ordinary differential equation ($D = d = 0$, ignoring the boundary conditions).
b) For D,$d > 0$ show that the equilibrium becomes unstable in the kth eigenmode spanned by $\sin \frac{k\pi}{L}$ when

$$b > 1 + \frac{a^2 D}{d} + \frac{\pi^2}{L^2} Dk^2 + \frac{d^2 L^2}{d\pi^2 k^2}$$

for some $k = 1, 2, 3, \ldots$
c) Hence, show that when $L = \pi k (Dd/a)^{1/4}$ and

$$d < D \quad \text{and} \quad a > 2\sqrt{\frac{D}{d}}\left(1 - \frac{D}{d}\right)^{-1},$$

the equilibrium becomes unstable in the kth mode, as b decreases below $1 + Da^2/d + 2a\sqrt{D/d}$.
d) Analyse the bifurcation structure in this case.

Exercise 2.8
The exothermic chemical reaction of bulk materials held in storage is a common hazard. Consider the distribution of temperature, u, within such a material, which is increased by the release of energy from the reaction. We have the model

$$u_t = D\Delta u + \alpha e^u, \qquad x \in \Omega, \ \ t > 0,$$

where D is the heat conductivity, u is the temperature distribution, and α is a positive constant. The rate term αe^u is a simplification of the more usual Arrhenius law.
Let us suppose that u is held fixed ($= u_0$) at the boundary of our domain, Ω.
Suppose $\Omega = [-L, L]$ in one dimension.
a) Consider the steady-state problem

$$Dw_{xx} + \alpha e^w = 0, \ \ w(-L) = w(L) = 0.$$

Draw the phase plane.
b) Show that $w = 2\ln\{e^{A/2}\text{sech}\,(Bx)\}$, with A any positive constant, and $B = (\alpha e^A/2D)^{1/2}$ solves the steady state equation.

c)Sketch the relationship between A and the ratio α/D given by

$$e^{A/2}\text{sech}\,(BL) = 1$$

where L is fixed.

d) Discuss the steady-state solutions. Show that for α/D large there are no solutions, while for α/D small there are two solutions. What is the behaviour of these solutions as $\alpha/D \to 0$?

What is the critical value of α/D? Use an iterative scheme, such as the Newton method, to find an approximate value.

3 Plane waves

3.1 Introduction

We have introduced plane waves in section 1.5. Simply, they are waves having a fixed profile, propagating in a fixed direction, with constant speed. In two or three dimensions, the wave solution is a function of a single travelling variable, say $z = \mathbf{n}.\mathbf{x} - ct$, where c is the wave speed and $\mathbf{n}$ is the unit vector denoting the direction of propagation. The wave is independent of any variations in $\mathbf{x}$ perpendicular to $\mathbf{n}$. Hence the term plane waves.

If we wish to find plane wave solutions for a given system, we may as well restrict $\mathbf{x}$ to be in $\mathbf{R}$, and consider a one dimensional problem, with the x-axis measuring distance in the direction of possible propagation.

Plane waves have a number of direct applications. In models of nerve axons, we are interested in the propagation of pulses of depolarization through (essentially one-dimensional) domains, [21]. In combustion theory, a plane wave describes how solid fuel or gas is burnt as the flame front passes through a long, narrow domain.

In such applications, the plane wave problem is the one of central interest (other issues, such as threshold-type behaviour, being secondary to the models' ability to support wave solutions in agreement with those observed as real phenomena). However, even when the waves themselves are not of fundamental interest, they may be of use in describing the evolution of solutions. On large (or infinite) domains, a solution to an initial-boundary value problem may well develop behaviour which is locally wave-like. If we are aware of the possible plane wave solutions, we can intuitively describe such behaviour in terms of waves sweeping over subdomains, colliding with one another, or meeting boundaries.

The analysis in Chapter 4 describes many exotic nonplanar wave phenomena. Here, the starting point is the given reaction-diffusion system and its possible plane wave solutions.

There is a vast literature concerning plane wave solutions for parabolic systems. This reflects not only the interest in their applications, but also the growth of qualitative analysis of solutions for nonlinear ordinary differential equations.

Consider the system

$$\mathbf{u}_t = D\mathbf{u}_{xx} + \mathbf{f}(\mathbf{u}, \mathbf{u}_x) \tag{3.1.1}$$

defined for $x \in \mathbf{R}$, $t \geq 0$. Here, $\mathbf{u} \in \mathbf{R}^m$, D is an $(m \times m)$ matrix, and $\mathbf{f}$ is some given (smooth) mapping. Setting $z = x - ct$, and $\mathbf{u} = \mathbf{w}(z)$, we seek a plane wave with profile given by w and speed c. We have

$$0 = D\mathbf{w}_{zz} + c\mathbf{w}_z + f(\mathbf{w}, \mathbf{w}_z). \tag{3.1.2}$$

Here, both c and w must be determined simultaneously. As $|z| \to \infty$, we may impose a certain behaviour upon w. For example, suppose $\mathbf{u}_1, \mathbf{u}_2 \in \mathbf{R}$ are such that $f(\mathbf{u}_i, 0) = 0$, $i = 1, 2$. Then $\mathbf{u}_1$ and $\mathbf{u}_2$ represent rest states for (3.1.1). We may seek waves connecting such equilibria by imposing

$$\begin{aligned} \mathbf{w} &\to \mathbf{u}_1 \text{ as } z \to \infty, \\ \mathbf{w} &\to \mathbf{u}_2 \text{ as } z \to -\infty. \end{aligned} \tag{3.1.3}$$

The natural space for solutions of (3.1.2) is $C^2(\mathbf{R})$, so that the derivatives present in (3.1.2) exist, and the wave remains uniformly bounded.

A solution $\mathbf{w}$ satisfying (3.1.2) and (3.1.3) can be thought of as an orbit connecting $\mathbf{u}_1$ and $\mathbf{u}_2$ in the so-called **phase space**. This is simply $(\mathbf{w}, \mathbf{w}_z)$-space $\approx \mathbf{R}^{2m}$. We may rewrite (3.1.2) as a first-order system of ordinary differential equations:

$$\begin{aligned} \mathbf{w}_z &= \mathbf{p}, \\ D\mathbf{p}_z &= -c\mathbf{p} - \mathbf{f}(\mathbf{w}, \mathbf{p}), \end{aligned} \tag{3.1.4}$$

and our orbit satisfying (3.1.2) leaves the rest point $(\mathbf{w}, \mathbf{p}) = (\mathbf{u}_2, 0)$ as z increases from $-\infty$, and approaches the rest point $(\mathbf{w}, \mathbf{p}) = (\mathbf{u}_1, 0)$ as $z \to \infty$.

If $\mathbf{u}_1 \neq \mathbf{u}_2$, such an orbit is called **heteroclinic**. The resulting wave changes the state of the system as it passes. The simplest example is the front solution developed in section 1.5.

If $\mathbf{u}_1 = \mathbf{u}_2$ in (3.1.3), then our orbit starts out from and finishes at the same rest point. In this case, the orbit is called **homoclinic**, and the resulting wave is a kind of **pulse**. We shall begin to analyse such waves in section 3.3.

Other alternatives suggest themselves; for example, we may seek periodic solutions for (3.1.4). The resulting wave is a travelling periodic wave (with appropriate speed, c).

The central point is that the nonlinear eigenvalue problem, (3.1.2), must be solved for appropriate $(\mathbf{w}, c) \in C^2(\mathbf{R}) \times \mathbf{R}$, and techniques must be developed to take care of the nonlinear nature of the system.

An important aspect of such problems is that the possession or otherwise of wave solutions is a **qualitative** property of systems such as (3.1.2). Suppose that $(\mathbf{w}, c) = (\mathbf{w}_0, c_0)$ satisfies (3.1.2-3), for example. Let us perturb the nonlinearity, $\mathbf{f}$, to become

$$\mathbf{f}^*(\mathbf{w}, \mathbf{w}_z) = \mathbf{f}(\mathbf{w}, \mathbf{w}_z) + \varepsilon \mathbf{g}(\mathbf{w}, \mathbf{w}_z),$$

where $\mathbf{g}(\mathbf{u}_i, 0) = 0$, $i = 1, 2 \ldots$, $\mathbf{g}$ is uniformly bounded, and $\varepsilon > 0$ is small. Let us seek a solution of

$$\begin{aligned} D\mathbf{w}_{zz} + c\mathbf{w} + \mathbf{f}^*(\mathbf{w}, \mathbf{w}_z) &= 0, \\ \mathbf{w} \to \mathbf{u}_1 \quad z &\to \infty, \\ \mathbf{w} \to \mathbf{u}_2 \quad z &\to -\infty. \end{aligned} \tag{3.1.5}$$

For $\varepsilon > 0$, we perturb $(\mathbf{w}_0, c_0)$ by setting

$$\begin{aligned} \mathbf{w} &= \mathbf{w}_0 + \varepsilon \mathbf{w}_1 + \varepsilon^2 \mathbf{w}_2 + \cdots, \\ c_0 &= c_0 + \varepsilon c_1 + \varepsilon^2 c_2 + \cdots \end{aligned} \tag{3.1.6}$$

Here, we must have

$$\mathbf{w}_i \to 0 \quad \text{as} \quad |z| \to \infty, \quad i = 1, 2, \ldots$$

Substituting (3.1.6) into (3.1.5), we have to $O(\varepsilon)$

$$\begin{aligned} D\mathbf{w}_{1zz} + c_0\mathbf{w}_{1z} + \mathbf{f}_{\mathbf{w}}(\mathbf{w}_0, \mathbf{w}_{0z})\mathbf{w}_1 & \\ + f_{\mathbf{w}_z}(\mathbf{w}_0, \mathbf{w}_{0z})\mathbf{w}_{1z} &= c_1\mathbf{w}_{0z} - \mathbf{g}(\mathbf{w}_0, \mathbf{w}_{0z}). \end{aligned} \tag{3.1.7}$$

Here, $\mathbf{f}_{\mathbf{w}}$ is the $(m \times m)$ matrix with (i, j)th element

$$\frac{\partial f_i}{\partial w_j}(\mathbf{w}_0, \mathbf{w}_{0z}).$$

Similarly, $f_{\mathbf{w}_z}$ is the linearization of $\mathbf{f}$ with respect to the $\mathbf{w}_z$ argument.

It is easy to check that $\mathbf{w}_{0z}$ is annihilated by the differential operator on the left-hand side of (3.1.7). Thus, the solution of (3.1.7) for $\mathbf{w}_1$ requires the imposition of a solvability condition (see Box C and Box H). Let $\mathbf{w}^*$ be the solution for the homogeneous adjoint problem:

$$D\mathbf{w}^*_{zz} - c_0\mathbf{w}^*_z + \mathbf{f}_{\mathbf{w}}(\mathbf{w}_0, \mathbf{w}_{0z})^T\mathbf{w}^* - \mathbf{f}_{\mathbf{w}_z}(\mathbf{w}_0, \mathbf{w}_{0z})^T\mathbf{w}^*_z = 0,$$

$$\mathbf{w}^* \to 0 \quad |z| \to \infty.$$

Then (3.1.7) may be solved for $\mathbf{w}_1$ providing

$$\int_{\mathbf{R}} \mathbf{w}^{*T}(c_1\mathbf{w}_{0z} - \mathbf{g}(\mathbf{w}_0, \mathbf{w}_{0z}))\, dz. \tag{3.1.8}$$

This fixes c_1, and $\mathbf{w}_1$ is unique up to the addition of linear multiples of $\mathbf{w}_{0z}$. These terms may be discarded so that

$$\int_{\mathbf{R}} \mathbf{w}_1^T . \mathbf{w}_{0z}\, dz = 0,$$

since such perturbations of $\mathbf{w}_0(z)$ can be annihilated by translating the z-variable (see section 3.2).

In singular cases where (3.1.8) cannot be solved for c_1, our perturbation theory will be invalid but, for the most part, we can always perturb the wave solution, $(\mathbf{w}_0, c_0)$, to obtain a new solution for the perturbed problem, (3.1.5).

In some applications, the models will be very precisely known (e.g. for chemical reactions, where rate constants and diffusivities may be accurately determined). In other situations, the models may be more **qualitative**, rather than **quantitative** (for example, the FitzHugh-Nagumo equations, see section 3.4). Here, the precise forms of the nonlinearities remain unknown, and they are assumed to have generic forms, described by schematic graphs. The front problem in section 1.5 was considered for any $f(u)$ of the type depicted in Figure 1.7. The degree of accuracy to which the terms are known does not affect certain basic properties of systems although, of course, we must make some definite choices as soon as numerical computations are to be undertaken.

In this chapter, we concentrate on methods which prove useful for such qualitative systems. In section 3.3, we introduce the notion of excitable systems, since these will play a central role in subsequent examples both here and in Chapter 4.

In sections 3.4 and 3.5, we utilize an asymptotic, transition layer, approach to travelling wave problems. Though relatively straightforward, this technique is of use in a wide variety of nonlinear systems.

In section 3.6, we introduce piecewise linear systems. Here, we replace the nonlinear terms with piecewise linear approximations. Then, we seek to solve the linear problems in subsets of the z-domain, applying continuity conditions at the appropriate points where switching occurs. Although this seems crude at first sight (and purists may throw up their hands in horror!), it is an approach that yields a large amount of information concerning the existence of waves and their dependence upon certain model parameters. Also, we must bear in mind that the qualitative behaviour of the piecewise linear system and that of the original nonlinear system are likely to be similar provided that the solution does not linger for too long close to our *artificial* switches or discontinuities.

Nonlinear systems of the form (3.1.4) are amenable to other, more qualitative, types of analysis. Techniques utilizing topological indexes and shooting methods, [10],[70], have been variously successful. In particular, [10] is recommended as an introduction to the qualitative theory of differential equations. However, we shall not stress these approaches in the present work, since this is a vast subject in its own right, and would divert us from our basic aim which is to stress **constructive** techniques.

3.2 Stability

Before considering further examples of travelling waves, we consider the important question of their stability. Simply speaking, all this requires is that we check that small perturbations to the wave die out, or result in a mere phase shift.

Consider the equation for $\mathbf{u}(x,t) \in \mathbf{R}^m$:

$$\mathbf{u}_t = D\mathbf{u}_{xx} + \mathbf{f}(\mathbf{u}), \qquad x \in \mathbf{R}, \quad t > 0. \tag{3.2.1}$$

Here, D is a diagonal matrix, and f is some smooth vector field. Substituting $\mathbf{u} = \Phi(x - ct)$ into (3.2.1), we have a travelling wave solution if and only if $(\Phi, c) \in C_2(\mathbf{R}) \times \mathbf{R}$, say, and satisfy

$$D\Phi_{zz}(z) + c\Phi_z(z) + \mathbf{f}(\Phi(z)) = 0. \tag{3.2.2}$$

Suppose, for the moment, that this is so. First, notice that $\Phi(z)$ is determined up to a translation in z. (In fact, we should really think of a travelling wave as a family of waves related via rigid translation.) Assume Φ is smooth; we differentiate (3.2.2) with respect to z, so that

$$D\Phi_{zzz} + c\Phi_{zz} + \mathbf{f}_\Phi(\Phi)\Phi_z = 0. \tag{3.2.3}$$

Here, $\mathbf{f}_\Phi(\Phi)$ is the $m \times m$ matrix, with (i,j)th element $\frac{\partial f_i}{\partial \Phi_j}$.

Now we consider the stability of Φ (and its translates). We set

$$\mathbf{u}(x,t) = \Phi(z) + \mathbf{w}(z,t), \tag{3.2.4}$$

where $\mathbf{w}$ is some small perturbation. We use the moving coordinate z, so that we may see how the perturbation evolves in the moving frame of reference associated with the wave; if $\mathbf{w}$ decays, it will do so with respect to both the z-frame and the x-frame.

Substituting (3.2.4) into (3.2.1) and retaining only linear terms in $\mathbf{w}$, we have

$$\mathbf{w}_t = D\mathbf{w}_{zz} + c\mathbf{w}_z + \mathbf{f}_\Phi(\Phi)\mathbf{w}, \qquad z \in \mathbf{R}, \quad t > 0. \tag{3.2.5}$$

Now we seek to apply the stability theory of section 1.4. Firstly, we assume $\mathbf{w}$ is in some suitable Banach space, X say, and define the operator A so that (3.2.5) may be written:

$$\mathbf{w}_t + A\mathbf{w} = 0, \qquad \mathbf{w}(0) \in X, \quad t > 0. \tag{3.2.6}$$

For example, we might choose $X = L_2(\mathbf{R}, \mathbf{R}^m)$, and define A firstly for C_2 functions in X, and then extend it to become a closed, densely defined operator on X.

Our equilibrium, $\mathbf{u} = \Phi$, corresponds to $\mathbf{w} = 0$ and is asymptotically stable if the spectrum of A lies strictly to the right of the imaginary axis in the complex plane.

However, if $\Phi_z \in X$, then (3.2.3) implies $A\Phi_z = 0$, so that zero is an eigenvalue for A. However, all is not lost since this particular eigenmode (spanned by Φ_z) is that generated by perturbations equivalent to infinitesimal translations of Φ, that is we have

$$\Phi(z+h) = \Phi(z) + h\Phi_z(z) + O(h^2).$$

We have already noticed that these translations result in a family of wave solutions. Hence, such perturbations may only result in a phase shift of the original wave. When this is the case, it makes sense to discount such perturbations and define the stability of Φ as the stability of the family of waves obtained by rigid translations of Φ. More precisely, we say that Φ is **stable** if and only if $\mathbf{u}(x,t) = \Phi(z) + \mathbf{w}(z,t)$ converges to some $\Phi(z+h)$ for some h constant and finite, as $t \to \infty$. This will be the case whenever zero is a simple eigenvalue of A and the remainder of the spectrum lies in a half-space $\{\lambda : Re(\lambda) \geq \beta\}$ for some real $\beta > 0$. This is an application of more general results concerning the stability of families of equilibria (see [35] Chapter 5, for example). In such circumstances, we have $\mathbf{w} \to 0$ as $t \to \infty$ in (3.2.6) provided the projection of $\mathbf{w}(0)$ on to the linear subspace spanned by Φ_z is zero. (If X is an Hilbert space, then $\mathbf{w} \to 0$ provided that $\mathbf{w}(0)$ is orthogonal to Φ_z.)

For coupled systems, the operator A is often difficult to analyse, and the determination of its spectral properties represents a substantial problem. However, we shall consider a few simple examples shortly.

As indicated in section 1.4, the choice of our underlying space X is crucial. For example, if we force functions in X to decay rapidly enough, we can push the spectrum to the right and even lose our eigenvalue at the origin (when Φ_z ceases to be in X). We should aim to keep X as large as possible so that we do not lose perturbations (such as small translations) which we feel are intuitively reasonable in the physical problem.

Example 1

We return to the example discussed in section 1.5.

We considered

$$u_t = u_{xx} + f(u), \quad x \in \mathbf{R}, \quad t > 0, \tag{3.2.7}$$

where f was of the form depicted in Figure 1.6. By way of example, we shall choose $f = u(1-u)(u-a)$ for some constant $a \in (0,1)$. We showed that there exists a unique speed c such that (3.2.7) has a monotonic wave-front solution of the form

$$u = \phi(z), \quad z = x - ct$$

where

$$\lim_{z\to\infty} \phi = 0$$

$$\lim_{z\to-\infty} \phi = 1$$

(let $z \to -z$ in (1.5.3)).

We shall show that this wave is stable in a general sense, following [35]. In the notation introduced earlier, we have

$$-Aw = w_{zz} + cw_z + f'(\phi)w.$$

Here, we shall choose the underlying space X to be $L_2(\mathbf{R})$ (although the argument would work just as well in $C(\mathbf{R})$).

Firstly, as $z \to +\infty$, we have $\phi \to 0$ and $f'(\phi) \to -a$. As $z \to -\infty$, $\phi \to 1$ and $f'(\phi) \to -1+a$. Now we employ the theorem in Box D, section 1.4, which ensures that the **essential spectrum** of A is bounded to the right of

$$Re(\lambda) = \min\{a, 1-a\}.$$

Recall that the spectrum of A is made up of isolated eigenvalues of finite multiplicity as well as the essential spectrum (which includes any continuous and residual spectrum).

Now consider the eigenvalues, λ, of A. Suppose $Re(\lambda) \leq 0$, and that there exists $v \in L_2$ satisfying

$$0 = -Av + \lambda v = v'' + cv' + f'(\phi)v + \lambda v.$$

As $z \to \infty$, v must decay to zero at least as $O(e^{-cz})$, so certainly

$$y(z) = v(z)e^{\frac{c}{2}z}$$

decays exponentially as $|z| \to \infty$. But now $y(z) \in L_2(\mathbf{R})$, and satisfies the self-adjoint problem

$$y_{zz} + \left(f'(\phi) + \lambda - \frac{c^2}{4}\right)y = 0, \quad z \in \mathbf{R}, \quad y \to 0, \quad |z| \to \infty. \tag{3.2.8}$$

Hence, all eigenvalues, λ, are real, and multiplying by y and integrating over $\mathbf{R}$, we have

$$\lambda \int y^2 dz = \int y_z^2 - \left(f'(\phi) - \frac{c^2}{4}\right)y^2 dz. \tag{3.2.9}$$

Now, when $\lambda = 0$, we observe that $y = e^{-\frac{c}{2}z}\phi_z(z)$ is a solution (since $A\phi_z = 0$, due to the translational invariance of the wave). Let $\psi(z) = e^{-\frac{c}{2}z}\phi_z(z)$, so that (3.2.9) may be rewritten as

$$\begin{aligned}\lambda \int y^2 dz &= \int y_z^2 + \frac{\psi_{zz}y^2}{\psi} dz \\ &= \int y_z^2 - \frac{2yy_z\psi_z}{\psi} + \frac{\psi_z^2 y^2}{\psi^2} dz \\ &= \int \psi^2 \left(\frac{d}{dz}\left(\frac{y}{\psi}\right)\right)^2 dz.\end{aligned}$$

Thus, $\lambda \geq 0$, and $\lambda = 0$ if and only if $y = \text{constant}.\psi$. However, this implies that $w = \text{constant}.\phi_z$, so $\lambda = 0$ is a simple eigenvalue of A as required. Hence, if we consider the initial-value problem for (3.2.7), with $u(x, 0) - \phi(x) \in L_2(\mathbf{R})$, and small enough, then there exists h, real, such that

$$\|u(x,t) - \phi(x - ct + h)\|_{L_2} \to 0 \quad \text{as} \quad t \to \infty.$$

The above analysis relied on converting the original problem to a self-adjoint form, and then the fact that ϕ_z (and hence ψ) was nonvanishing.

Example 2

The nonlinearity may depend on spatial derivatives of the state variables. Similar considerations to those above apply. Consider

$$u_t = u_{xx} - uu_x.$$

Here, we have travelling waves of the form

$$u = \phi(z) = c - a\tanh(az/2)$$

for any constants $a > 0$, and wave speed, c. (Here, as usual, $z = x - ct$.) Fixing a and c, we have, in the notation introduced above,

$$Aw = -w_{zz} + (\phi(z) - c)w_z + \phi_z w. \tag{3.2.10}$$

By employing the analysis of Box D, we may see that the **essential spectrum** of A lies on and to the right of the parabola

$$Re(\lambda) = |a|^2 Im(\lambda)^2.$$

However, by weighting the function space by demanding that $w \in X$ if and only if

$$\|w\|_X^2 = \int w^2 \cosh^2 \frac{\gamma z}{2} dz < \infty$$

for some $\gamma \in (0, 2a)$ fixed, we may push the essential spectrum to the right of the imaginary axis (see Box D, section 1.4). In fact, we may check that the essential spectrum is strictly to the right of the line

$$Re(\lambda) = \frac{a\gamma}{2} - \frac{\gamma^2}{4} \ > \ 0.$$

The decay of $\phi_z(z)$ is $O(e^{-|z|a})$ as $|z| \to \infty$, so $\phi_z \in X$ provided $\gamma < a/2$.

If we choose $\gamma = a$, then for each $w \in X$, we have $u = w \cosh \frac{a}{2} z$ in $L_2(\mathbf{R})$. Hence, substituting for w in (2.3.10), we see that if w is an eigenfunction for A in X, then u must satisfy

$$-\lambda u = u_{zz} + \frac{a^2}{2}\left(\text{sech}^2(az/2) - \frac{1}{2}\right)u$$

which is a self-adjoint problem in $L_2(\mathbf{R})$. Now, an argument similar to that in Example 1 above will show that $\lambda = 0$ is a simple eigenvalue for both problems and that there are no eigenvalues with $Re(\lambda) \ < \ 0$.

Thus, the wave $\phi(z)$ is stable with respect to perturbations in our weighted space X.

3.3 Excitable systems

In many of the applications of reaction-diffusion systems, interest is centred upon so-called **excitable systems** and their wave-like solutions.

Excitability may be thought of as a property of the reaction kinetics, and is defined as follows.

Consider a system of ordinary differential equations

$$\mathbf{u}_t = \mathbf{F}(\mathbf{u}), \tag{3.3.1}$$

where $\mathbf{u} = 0$ is a locally stable equilibrium (the matrix $\{\partial F_i/\partial u_j\}$ possesses eigenvalues with negative real parts).

By definition, small perturbations to the equilibrium will decay (exponentially) as $t \to \infty$. Larger perturbations may behave differently, depending upon the nonlinear terms within $\mathbf{F}$.

The system (3.3.1) is **excitable** if there are intermediate sized perturbations of the resting equilibrium, which result in time-dependent solutions taking a prolonged excursion through the state space before returning to rest. The divide between the two qualitatively different kinds of behaviour is known as a **threshold** effect. Sub-threshold perturbations (or excitations) result in a simple return to the quiescent equilibrium, $\mathbf{u} = 0$, while super-threshold perturbations (or excitations) result in trajectories which sweep the state towards some new (excited) regime, before returning back (perhaps via a prolonged cycle of behaviour) towards the quiescent equilibrium.

A simple example will suffice to illustrate these ideas. Consider

$$\begin{aligned} u_t &= f(u) - v, \\ v_t &= \varepsilon(u - v), \end{aligned} \tag{3.3.2}$$

where $f = u(u-a)(1-u)$, for $a \in (0, 1/2)$. Here, $1 >> \varepsilon > 0$, so that v_t is generally small compared to u_t. Hence, v is called the **slow** variable, and only dominates the behaviour of the system while $f(u) \approx u$, but even then v_t is small, and v moves slowly.

The u-variable is called the **fast** variable. When $f(u) \neq v$, the fast equation in (3.3.2) dominates the behaviour of the system until such time as both u_t and v_t are both $O(\varepsilon)$. In Figure 3.1, we sketch the nullclines for (3.3.2), along with a few trajectories. The rest point, $(u, v) = (0, 0)$, is locally stable (check that the linearized system possesses

$$\lambda = \frac{-(a+\varepsilon) \pm \sqrt{(a-\varepsilon)^2 - 4\varepsilon}}{2}$$

as its eigenvalues). Small perturbations result in trajectories such as those starting out at A or B, which return quickly to equilibrium. Larger perturbations such as those starting from C or D result in much more distinct behaviour. These trajectories move up to the right-hand branch of f (the excited regime) before being shifted back towards the left-hand branch, and returning to rest. The near-horizontal parts of the trajectory take place on the fast time scale. Figure 3.2 shows the time-dependent profiles of u and v for the trajectory starting out at C.

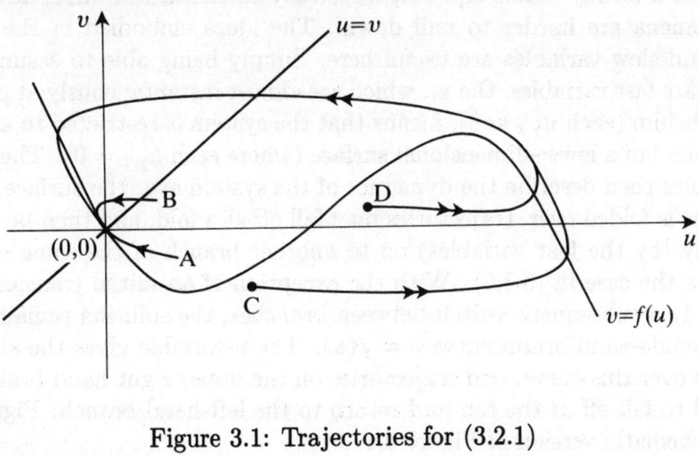

Figure 3.1: Trajectories for (3.2.1)

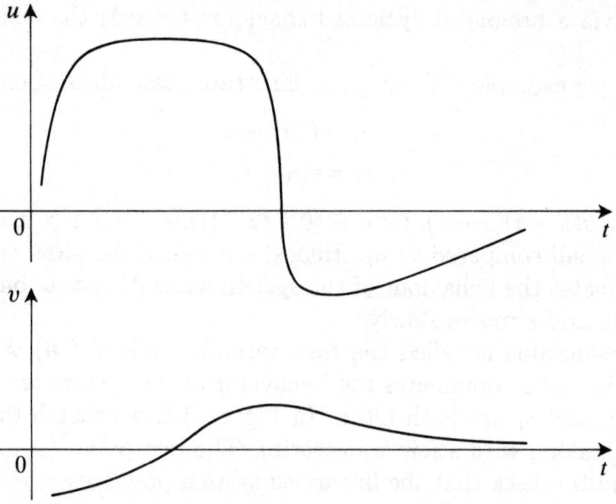

Figure 3.2: Time profiles for a solution of (3.2.1)

Other excitable systems may be more difficult to classify. The requirement of a locally stable equilibrium is easy to check, but **threshold**-type phenomena are harder to nail down. The ideas embodied in the use of fast and slow variables are useful here. Simply being able to assume that there are fast variables, the u_i, which are almost instantaneously at pseudo-equilibrium (each $u_{j\ t} \approx 0$), means that the system is restricted to a neighbourhood of a lower-dimensional surface (where each $u_{j\ t} = 0$). The slower variables then describe the dynamics of the system over the surface. If the surface is folded over, trajectories may fall off at a fold, and then be shifted rapidly (by the fast variables) on to another branch of the same surface. This is the case in (3.3.2). With the exception of an initial transient, and possibly a subsequent switch between branches, the solution remains near the pseudo-equilibrium curve $v = f(u)$. The v-variable gives the slow dynamic over this curve, and trajectories on the upper right-hand branch are forced to fall off at the top and return to the left-hand branch. Figure 3.3 is a schematic version of Figure 3.1.

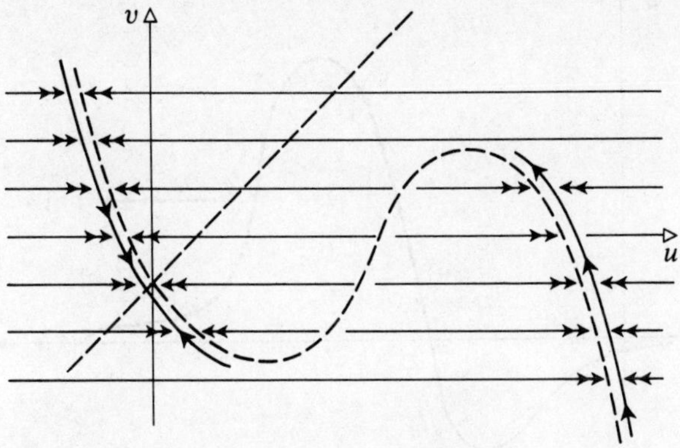

Figure 3.3: Fast and slow dynamics.

Clearly, for any given system, if there are any small parameters available, and a consequent fast-slow classification of the state variables, it is much easier to look for threshold behaviour and excitability. The fast-slow construction is equivalent to the techniques of **matched asymptotic expansions** (see Box H), and **two-timing** in asymptotic analysis, [52].

When diffusion terms are added to (3.3.1), it is natural to look for wave solutions. The idea is that as the wave passes through a point, the behaviour of the state is similar to the time course of trajectories like those shown in Figure 3.2. The initial excitation is supplied by diffusive interacton with neighbouring points already excited by the earlier arrival of the wave. Thus, the wave propagates by successive neighbourhoods stimulating one another (via the diffusion in the fast variables) to become superthreshold, whereupon the dynamics begin to dominate. In one dimension, we expect that a system like (3.3.2) may well possess a travelling pulse solution, like that shown in Figure 3.4, moving with some constant speed, c.

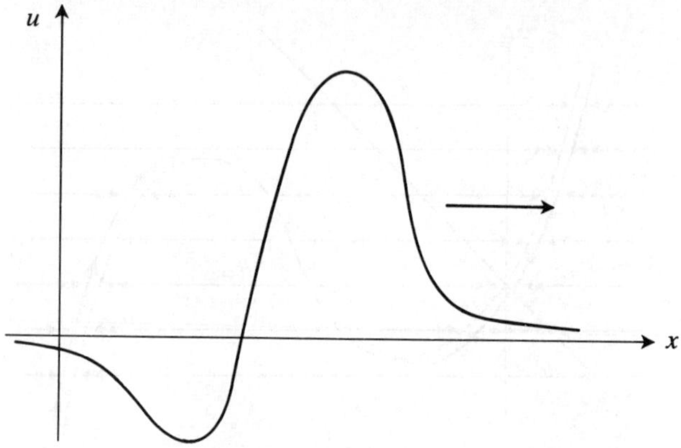

Figure 3.4: Travelling pulse.

Unfortunately, if we simply introduce a travelling variable ($z = x - ct$) and look for such a solution, we are forced to analyse a nonlinear first-order system of ordinary differential equations defined on $\mathbf{R}^3$ ((u, u_z, v)-space). This is not a straightforward exercise, particularly as the wave speed, c, is to be determined simultaneously with the wave profile.

Here, however, the idea of fast and slow variables comes into its own once more. In the next section, we shall show how one may seek waves with **transition layers**. As in Chapter 2, section 2.3, we obtain qualitative information about the solutions (and, of course, the wave speeds) without solving the full system.

3.4 Waves with transition layers

By exploiting any small parameters present in a system, we hope to construct travelling wave solutions by assuming the existence of suitable internal transition layers. As in section 2.3, this means that, away from such layers, we may solve a system of reduced order, whilst the behaviour of the layer profile must be obtained by a suitable rescaling of the independent travelling variable.

By way of illustration, we shall fix our attention upon the excitable dynamics introduced in section 3.3, equations (3.3.2). We introduce a diffusion term for the fast variable, u, and, for simplicity, we shall assume that any diffusion of the slow v-variable is negligible compared to the other processes operating.

We have met this system before in Chapter 1. It is usually known as the **FitzHugh-Nagumo** model for the conduction of nerve impulses along unmyelinated nerve axons, [65]. We have

$$\begin{aligned} u_t &= \varepsilon^2 u_{xx} + f(u) - v, \\ v_t &= \varepsilon(u - v), \end{aligned} \qquad x \in \mathbf{R}, \quad t \geq 0. \tag{3.4.1}$$

Here, we have rescaled the x-variable so that the diffusivity is ε^2, where $\varepsilon > 0$ is very small. The reason for this will become apparent shortly.

We shall assume that f is cubic:

$$f(u) = u(u - a)(1 - u) \quad \text{for} \quad a \in (0, 1/2),$$

although any function with the same qualitative form would do just as well.

In order to proceed further, let us rescale time by introducing the long time scale $\tau = \varepsilon t$. Then, writing $(u, v) = (u(x, \tau), v(x, \tau))$, (3.4.1) becomes

$$\begin{aligned} \varepsilon u_\tau &= \varepsilon^2 u_{xx} + f(u) - v, \\ v_\tau &= u - v, \end{aligned} \qquad x \in \mathbf{R}, \quad \tau > 0. \tag{3.4.2}$$

Next, we introduce the travelling variable, $z = x - c\tau$, and seek solutions $u = U(z)$, $v = V(z)$, representing a wave moving to the right with speed c. In view of our comments in the last section, we shall look for a pulse solution, as depicted in Figure 3.4. We have

$$\begin{aligned} \varepsilon^2 U_{zz} + \varepsilon c U_z + f(U) - V &= 0, \\ c V_z + U - V &= 0. \end{aligned} \tag{3.4.3}$$

We shall assume $c = O(\varepsilon^0)$ or smaller.

Firstly, note that since $1 >> \varepsilon > 0$, the first equation of (3.4.3) is in equilibrium provided that $V \approx F(U)$. Thus, we expect that solution trajectories will stay close to this curve except perhaps near to one or two points where they may jump from branch to branch (U_{zz} and U_z being large in such events). Specifically, we seek a pulse solution with the schematic form depicted in Figure 3.5. This solution has two transition layers, one at $z = 0$, where the fast variable, u, shifts rapidly up to the excited regime, and one at $z = z_1 < 0$, where the fast variable returns back to the lower branch of f before equilibrium is restored as $z \to -\infty$.

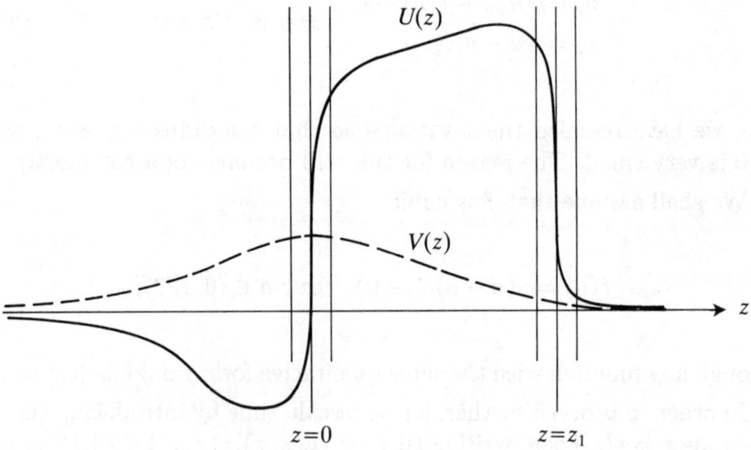

Figure 3.5: The singular pulse solution.

The up-jump at the forward transition layer (at $z = 0$) is known as the **wave front**, whilst the down-jump (at $z = z_1$) is called the **wave back**.

Both move with speed $c \geq 0$ in time to the right with respect to the x-frame.

To construct our singular pulse solution, we must specify the solution away from the transition layers, and then construct the transition profiles, determining the positive wave speed, c, and z_1 in the process.

Away from neighbourhoods $z = 0$ and z_1, from (3.4.3) we have

$$V = f(U),$$
$$cV_z + U - V = 0,$$

since ε is small and U_z, U_{zz} are assumed to be bounded.

For $z > 0$, or $z = z < z_1$, we assume $V = f(U)$ is equivalent to $U = h_-(V)$, where this denotes the left-hand branch of f (see Figure 3.6). For $z \in (z_1, 0)$, U is in the excitable regime and $V = f(U)$ is equivalent to $U = h_+(V)$, denoting the upper right-hand branch of f (again, see Figure 3.6).

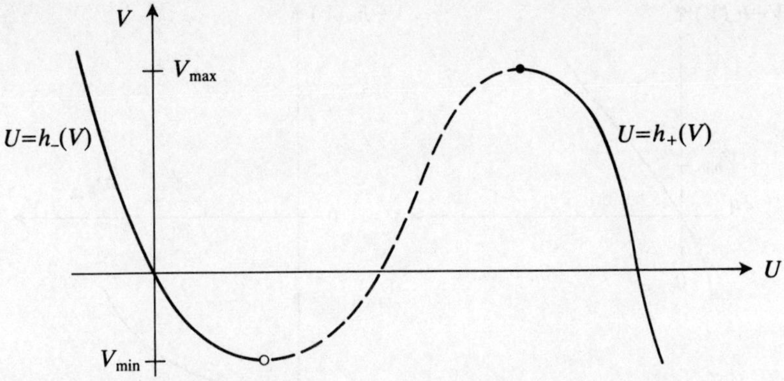

Figure 3.6: $h_\pm(V) = U$, pseudo inverses for $V = f(U)$.

Thus, for $z > 0$, we have

$$\begin{aligned} U &= h_-(V), \\ cV_z &= V - h_-(V), \\ \lim_{z\to\infty} V &= 0, \\ V(0) &= V_0, \text{ say.} \end{aligned} \tag{3.4.4}$$

For $z < z_1$, we have

$$\begin{aligned} U &= h_-(V), \\ cV_z &= V - h_-(V), \\ \lim_{z\to-\infty} V &= 0, \\ V(z_1) &= V_1, \text{ say.} \end{aligned} \tag{3.4.5}$$

Here, V_0 and V_1 are some constants still to be determined.

For $z \in (z_1, 0)$, we have

$$\begin{aligned} U &= h_+(V), \\ cV_z &= V - h_+(V), \\ V(z_1) &= V_1, \\ V(0) &= V_0. \end{aligned} \tag{3.4.6}$$

In Figure 3.7, we have drawn the nonlinearities associated with the differential equations in (3.4.4-6).

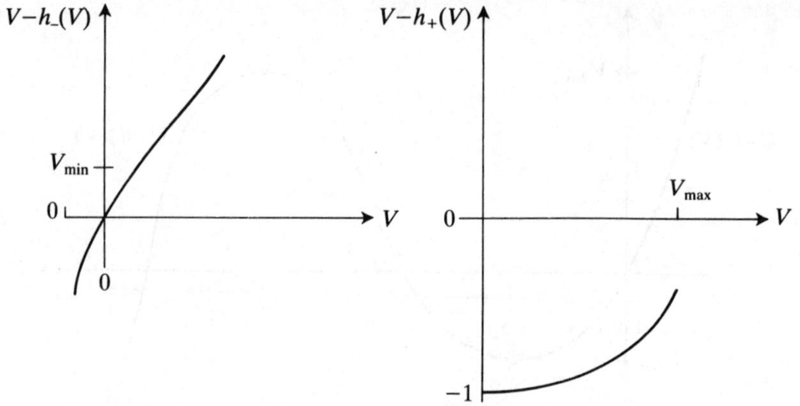

Figure 3.7: The nonlinearities in (3.4.4-6).

In (3.3.4), we require $V \to 0$ as $z \to \infty$, but $V = 0$ is an unstable equilibrium, so we must choose

$$V \equiv 0, \quad z > 0, \quad \text{and} \quad V(0) = V_0 = 0.$$

In (3.4.6), $cV_z < 0$. So, if $V(z_1) = V_1 \in (0, V_{max})$, there is a smooth solution satisfying the boundary conditions. Of course, V_1 fixes z_1 (and vice versa), so that z_1 decreases from zero as V_1 is increased from zero ($=V_0$).

In (3.4.5), $V = 0$ is an unstable equilibrium. So, for any V_1 fixed in $(0, V_{max})$, it is possible to find a solution on the half-line $z < z_1$, satisfying $V \to 0$, as $z \to -\infty$, together with $V(z_1) = z_1$.

Our outer solutions are complete. Once c and V_1 are given ($c \geq 0, V_1 \in (0, V_{max})$), the outer profiles and z_1 may be determined.

Next, we shall show that c is fixed by the **front** layer, at $z = 0$, while V_1 is fixed by the **back**, at $z = z_1$.

Consider a neighbourhood of $z = 0$. The V-variable has been matched at $z = 0$, whilst the U-variable must jump from $U = h_-(V_0) = 0$ ahead of the front to $U = h_+(V_0) = 1$ behind it. Introducing the scaled independent variable, $\varepsilon\xi = z$, and setting $U = \phi(\xi)$, $V = \psi(\xi)$ near $z = 0$, (3.4.3) becomes

$$0 = \phi_{\xi\xi} + c\phi_\xi + f(\phi) - \psi, \tag{3.4.7}$$
$$\phi \to 0 \text{ as } \xi \to \infty,$$
$$\phi \to 1 \text{ as } \xi \to -\infty.$$
$$0 = c\psi_\xi + \varepsilon(\phi - \psi), \tag{3.4.8}$$
$$\psi \to V_0 = 0 \text{ as } |\xi| \to \infty.$$

In the limit $\varepsilon \to 0$, it is the $\phi - \psi$ term in the second equation that is negligible on the new ξ-scale.

The required solution of (3.4.8) is simply $\psi \equiv 0$; so (3.4.7) reduces to the front problem discussed in section 1.5. (Compare (3.4.7) with (1.5.3).) There, we showed that this problem has a unique solution for (ϕ, c), and, in fact, since f is the cubic, we have an explicit solution

$$\phi(\xi) = \left(1 + \exp(\xi/\sqrt{2})\right)^{-1}$$
$$c = \sqrt{2}(1/2 - a).$$

Thus, $c > 0$, as required, since $a \in (0, 1/2)$. The requirement that (3.4.7) yields the transitional profile is enough to fix c.

Now consider the back at $z = z_1$. Let us set $\varepsilon\xi = (z - z_1)$ and write $U = \phi(\xi)$, $V = \psi(\xi)$ near $z = z_1$. (3.4.3) becomes

$$0 = \phi_{\xi\xi} + c\phi_\xi + f(\phi) - \psi, \tag{3.4.9}$$
$$\phi \to h_+(V_1) \text{ as } \xi \to \infty,$$
$$\phi \to h_-(V_1) \text{ as } \xi \to -\infty.$$
$$0 = -c\psi_\xi + \varepsilon(\phi - \psi), \tag{3.4.10}$$
$$\psi \to V_1 \text{ as } |\xi| \to \infty.$$

As $\varepsilon \to 0$, (3.4.10) yields $\psi \equiv V_1$ so that (3.4.9) becomes

$$0 = \phi_{\xi\xi} + \acute{c}\phi_\xi + f(\phi) - V_1, \tag{3.4.11}$$
$$\phi \to h_+(V_1) \text{ as } \xi \to \infty,$$
$$\phi \to h_-(V_1) \text{ as } \xi \to -\infty.$$

Consider a more general problem:

$$\begin{aligned} 0 &= \chi_{\zeta\zeta} + \gamma\chi_\zeta + f(\chi) - \alpha, \\ \chi &\to h_-(\alpha) \text{ as } \zeta \to \infty, \\ \chi &\to h_+(\alpha) \text{ as } \zeta \to -\infty. \end{aligned} \tag{3.4.12}$$

Here, α is fixed within the range $(V_{\min}, V_{\max})$ so that f has three roots, $h_-(\alpha)$, $h_+(\alpha)$, and one in between.

If $\gamma > 0$, a solution of (3.4.12) corresponds to a rightward propagating wave-front profile when $\xi = \zeta$ and $c = \gamma$. For example, (3.4.7) is of this form when $\psi \equiv 0$, and it was (3.4.12) which determined c ($= \gamma$, when $\alpha = 0$). If (3.4.12) has a solution with $\gamma \leq 0$, then by setting $\xi = -\zeta$ and $c = -\gamma$, we obtain a rightward propagating wave back ($\chi \to h_-(\alpha)$ as $\xi \to -\infty$ and $\chi \to h_+(\alpha)$, as $\xi \to +\infty$).

Hence, all possible wave fronts and backs for our transition profile problems can be obtained by examining the solutions of (3.4.12). We may solve (3.4.12) explicitly in the case where f is the cubic $u(u-a)(1-u)$. This is left as an exercise below. The important point is that the relationship between α and γ for solutions of (3.4.12) is of the form shown in Figure 3.8.

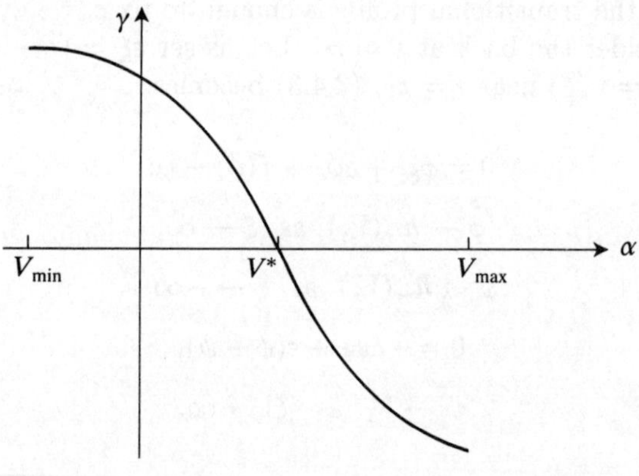

Figure 3.8: $\gamma - \alpha$ relationship for solutions of (3.4.12).

The value of $\alpha = V^*$ is that for which $\gamma = 0$, and here we have

$$\int_{h_-(\alpha)}^{h_+(\alpha)} (f(\chi) - \alpha)d\chi = 0.$$

(Multiply the second-order equation by χ_ζ and integrate from $-\infty$ to $+\infty$.)

Since f is cubic in our example, it turns out that γ in Figure 3.8 is an odd function of $(\alpha - V^*)$. For more general nonlinearities, f, this will not be so; nevertheless, $\gamma(\alpha)$ will retain the same qualitative behaviour.

Now we may solve (3.4.11). Set $\gamma = -c$ and read off the corresponding value of α from figure 3.8. Then we choose $V_1 = \alpha$, so that (3.4.11) possesses the required wave-back solution. Note that, by symmetry in figure 3.8 (and the fact that $\alpha = 0$ implies $\gamma = c$), we are assured that V_1 can be found.

We sketch our pulse solution for (3.4.1) schematically in Figure 3.9.

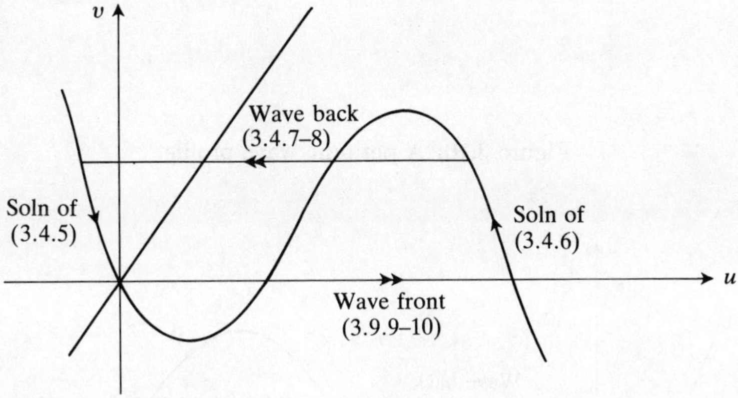

Figure 3.9: The pulse solution in the state space.

Our description of the pulse solution is now complete. Once V_1 is known, (3.4.5) determines z_1 as required.

Before leaving this problem, we can easily extend our transition layer analysis to construct **periodic wave** solutions for (3.4.1). For each $\alpha_1 \in (0, V^*)$, there is a point (α_1, γ_1) on the curve in Figure 3.8. There also exists $\alpha_2 \in (V^*, V_1)$ such that $(\alpha_2, -\gamma_1)$ is on the same curve. Thus, if we set $c = \gamma_1$, we have a wave front (when $V = \alpha_1$) and a wave back (when $V = \alpha_2$) both with speed c. As before, the slow equations

$$cV_2 + h_{\pm}(V) - V = 0$$

control the behaviour of the wave between the transition layers. It is straightforward to construct a periodic wave by utilizing successive backs and fronts for a fixed $\gamma_1 = c$, corresponding to a given α_1 in $(0, V^*)$. Figures 3.10 and 3.11 depict the wave profile and a schematic view of the periodic solution in (u, v)-space.

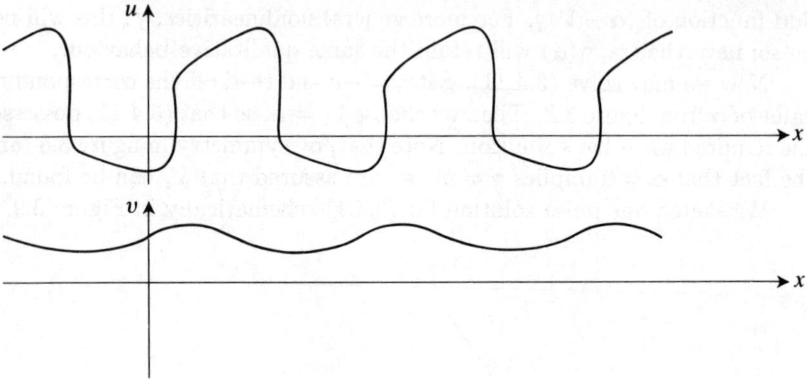

Figure 3.10: A periodic wave profile.

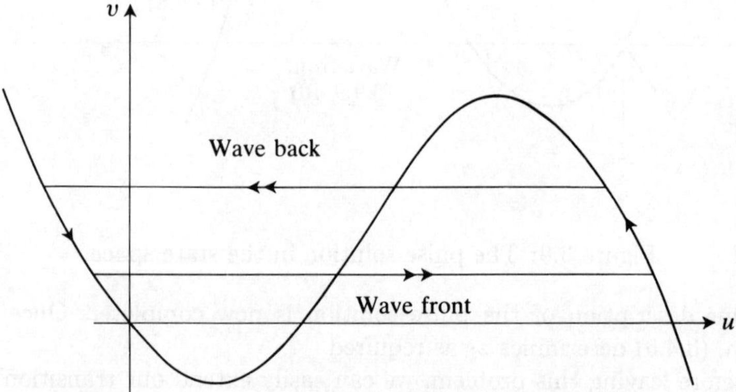

Figure 3.11: The periodic wave in state space.

Clearly, such waves exist for each choice of $\alpha_1 \in (0, V^*)$. Moreover, as $\alpha_1 \to 0$, the wave speed tends to that of the pulse solution obtained earlier. The **wavelength** of the periodic wave (i.e. the distance, on the x-scale, between successive wave fronts) must tend to infinity. To see this, we may estimate the wavelength from the outer solutions, since, as $\varepsilon \to 0$, the fronts and backs become instantaneous. We have, to $O(\varepsilon)$,

$$\lambda = \int_{\alpha_1}^{\alpha_2} \frac{c\,dV}{V - h_+(V)} + \int_{\alpha_2}^{\alpha_1} \frac{c\,dV}{V - h_-(V)}, \qquad (3.4.13)$$

(where λ denotes the wavelength). As $\alpha_1 \to 0$, the second integral in (3.4.13) tends to infinity since $V - h_-(V)$ has a simple zero at zero.

As $\alpha_1 \to V^*$, we have $\gamma_1 (= c) \to 0$, and $\alpha_2 \to V^*$ from above. Thus both terms in (3.4.13) tend to zero.

By allowing $\alpha_1 \in (0, V^*)$ to vary, we may plot the wave speed, c, of the consequent periodic solution, against the wavelength, λ. This is called the **dispersion curve**. Using the estimate (3.4.13) we obtain Figure 3.12.

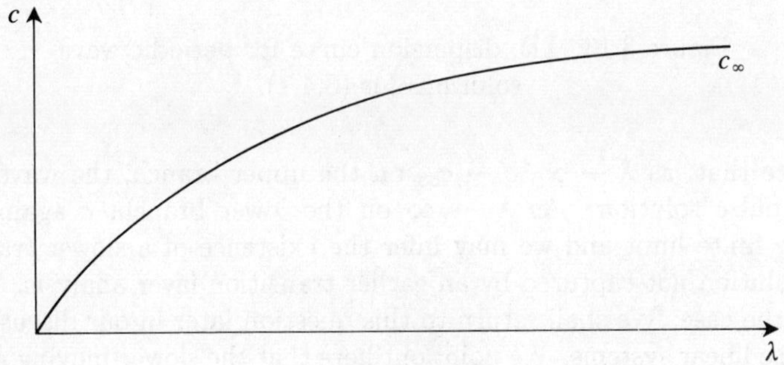

Figure 3.12: The dispersion curve for (3.4.1), using (3.4.13).

In fact, this is not quite the end of the story concerning periodic waves. As we noted earlier, as $c \to 0$ in Figure 3.12, we have $\lambda \to 0$, so the transitional fronts and backs become arbitrarily close together. However, this means that the underlying assumptions regarding the existence of transition layers are violated (there is no room to fit them in between outer solutions). Hence, we expect Figure 3.12 will become invalid for $\lambda = O(\varepsilon)$. The true dispersion curve for this problem is depicted in Figure 3.13. The lower branch of waves cannot be found using the above ideas but, in fact, turn out to be unstable and of little importance.

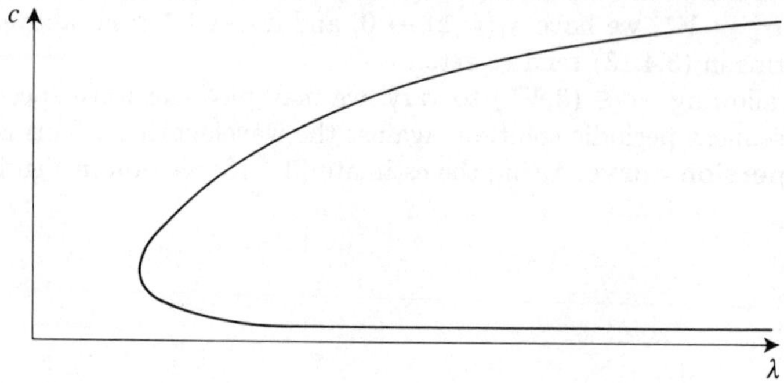

Figure 3.13: The dispersion curve for periodic wave solutions for (3.4.1).

Note that, as $\lambda \to \infty$, $c \to c_\infty$ on the upper branch, the wave speed of our pulse solution. As $\lambda \to \infty$ on the lower branch, c again tends to some finite limit and we may infer the existence of a slower travelling pulse solution not captured by an earlier transition layer analysis. This is indeed the case. We shall return to this question later in our discussion of piecewise linear systems. We point out here that the slower-moving pulse is unstable, while the faster one is stable. Although these stability results rest on considerations similar to those of section 3.2, they are far from trivial. In particular, the stability of the fast pulse solution remained a challenging open problem for a number of years.

Transition layers have been utilized in a variety of applications including waves in population biology (for models such as the time-dependent version of (2.3.1-2),[11]). As a further example, in the present section, we shall consider a wave-front problem arising in geochemistry. Here, the problem involves the flow of ground water through porous (fractured) bedrock, in this case, granite. The water contains dissolved oxygen molecules which react with a reducing agent, ferrous sulphide, which is fixed within the rock. The product also remains insoluble within the rock. The unoxidized rock has a different porosity to the reduced rock, but this will not play any role in our simple one-dimensional model. The problem is to determine how the front dividing oxidized and reduced rock propagates and, in particular, to estimate the speed of propagation.

Let $u(x,t)$ denote the concentration of dissolved oxygen within the water. Let $r(x,t)$ denote the concentration of the immobile reducing agent,

(FeS_2). Suppose that water flows from left to right with velocity v. The situation is depicted in Figure 3.14.

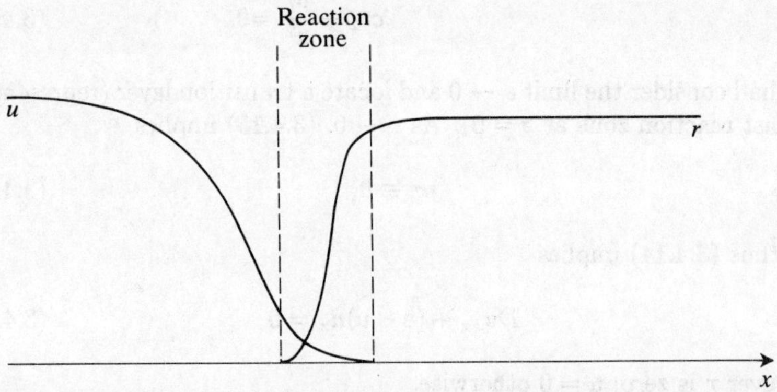

Figure 3.14: The convection of oxygen through rock.

To the left of the reaction zone, we have $r \approx 0$, and dissolved oxygen is convected by the water flow to the right. To the right of the reaction zone, we have $r \approx r_\infty$, the unoxidized concentration, and $u \approx 0$, since the local medium is strongly reducing. The width of the reaction zone is shown as small in Figure 3.14, since the reaction is assumed to be very fast compared to the time-scales involving the diffusion and convection of oxygen.

We have the reaction-diffusion-convection equations

$$\begin{aligned} u_t =& Du_{xx} - [vu]_x - \frac{ur}{\varepsilon^2} \\ r_t =& -\frac{ur}{\varepsilon^2}. \end{aligned}$$

Here, v is the velocity of the water, and we shall assume this to be constant. The reaction term represents the reduction of free oxygen and we wish to consider the model in the limit $\varepsilon \to 0$.

We shall assume $x \in \mathbf{R}$ and

$$\begin{aligned} u \to c_\infty,\ r \to 0 \quad &\text{as} \quad x \to -\infty, \\ u \to 0,\ r \to r_\infty \quad &\text{as} \quad x \to +\infty. \end{aligned}$$

Let us put $z = x - ct$ and write $u = u(z)$, $r = r(z)$ in order to seek a travelling wave profile moving rightwards with speed c. We have

$$\begin{aligned} Du_{zz} - vu_z + cu_z - \frac{ur}{\varepsilon^2} &= 0 \qquad (3.4.14)\\ cr_z - \frac{ur}{\varepsilon^2} &= 0. \qquad (3.4.15)\end{aligned}$$

We shall consider the limit $\varepsilon \to 0$ and locate a transition layer (representing the fast reaction zone at $z = 0$). As $\varepsilon \to 0$, (3.4.15) implies

$$ur \equiv 0, \qquad (3.4.16)$$

and thus (3.4.14) implies

$$Du_{zz} + (c - v)u_z = 0 \qquad (3.4.17)$$

wherever r is zero; $u = 0$ otherwise.

We solve (3.4.16-17) for the outer solutions satisfying the *boundary* conditions at $\pm\infty$. We have

$$z \geq 0; \quad u \equiv 0, \quad r \equiv r_\infty,$$

$$z \leq 0; \quad u = u_\infty \left(1 - e^{z(v-c)/D}\right), \quad r \equiv 0.$$

Here, we assume for the moment that $c < v$ (the front propagates at a slower rate than the water speed). We justify this *a posteriori*. We have also imposed the boundary condition $u = 0$ at $z = 0$. If this were not the case, the diffusion term in (3.4.14) would become the only dominant term inside the transition layer since the gradient in u would be arbitrarily large as $\varepsilon \to 0$.

Near $z = 0$, we must rescale the independent variable, and also u, since this itself is $O(\varepsilon)$ within a distance of $O(\varepsilon)$ to the left of the reaction zone. Set $\xi = z/\varepsilon$, $u = \varepsilon U(\xi)$, and $r = R(\xi)$. From (3.4.14-15), we obtain

$$\begin{aligned} DU_{\xi\xi} - UR + \varepsilon(c - v)U_\xi &= 0,\\ cR_\xi - UR &= 0. \end{aligned} \qquad (3.4.18)$$

As $\xi \to \infty$, $z > 0$ so we match U and R with the values of the outer solution

$$U \to 0, \quad R \to r_\infty \quad as \quad \xi \to \infty.$$

As $\xi \to -\infty$, we have $R \to 0$, while the matching for U is less trivial. The outer solution $u(z)(z \leq 0)$ approaches $u = 0$ (at $z = 0$) with nonzero slope. So we match the slopes (since $u = \varepsilon U$ is, by definition, already close to zero within the transition layer).

We have the equalities

$$U_\xi = \frac{u_\xi}{\varepsilon} = u_z$$

from our change of variables. Thus, we match

$$U_\xi \to \lim_{z\to 0^-} u_z$$

as $\xi \to -\infty$. The situation is depicted in Figure 3.15.

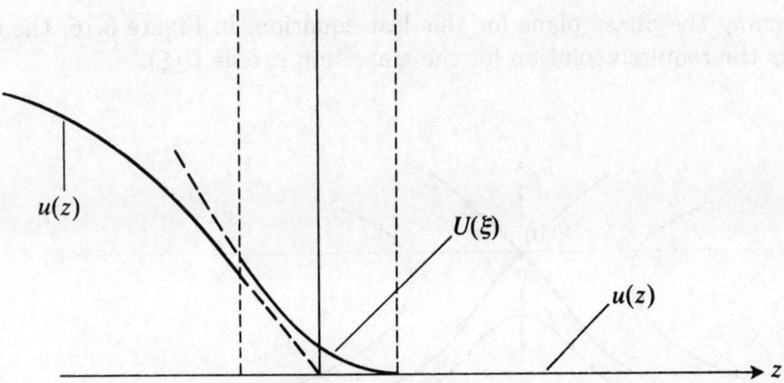

Figure 3.15: Inner and outer solutions for u.

Ignoring the $O(\varepsilon)$ term, we solve (3.4.18) as follows. Firstly note that

$$Du_{\xi\xi} - cR_\xi = 0$$

so that

$$Du_\xi - cR = \text{constant}.$$

By allowing $\xi \to \pm\infty$, we obtain

$$cr_\infty = -Du_z|_{z=0^-} \tag{3.4.19}$$

(where we have used the matching conditions). This is simply *continuity of flux of oxygen*, across the reaction zone. From our outer solutions, we obtain from (3.4.19)

$$cr_\infty = (v - c)u_\infty.$$

Hence

$$c = \frac{vu_\infty}{u_\infty + r_\infty},$$

which is the desired estimate of the speed of propagation of the wave front (notice that $c < v$ as required earlier).

In fact, we can analyse the front transition profile by using

$$R(\xi) = \frac{Du_\xi}{c} + r_\infty.$$

Thus, u in (3.4.18) satisfies

$$Du_{\xi\xi} - u(r_\infty + \frac{Du_\xi}{c}) = 0 \tag{3.4.20}$$

sketching the phase plane for this last equation, in Figure 3.16, the orbit A0 is the required solution for the transition profile $U(\xi)$.

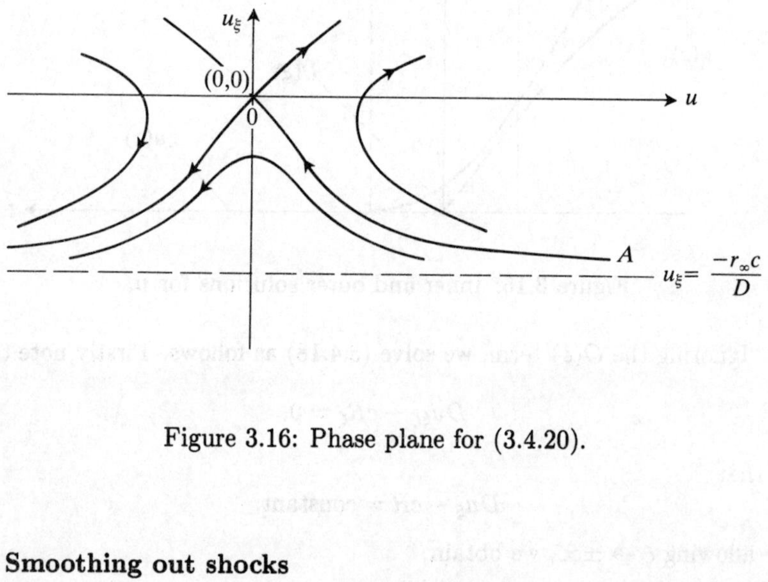

Figure 3.16: Phase plane for (3.4.20).

3.5 Smoothing out shocks

The use of transition layers in models having small diffusion coefficients enables us to take a sideways view at some of the problems related to **shock waves** in quasi-linear hyperbolic systems.

In many physical applications, models are based on conservation laws of the form

$$\mathbf{u}_t + \nabla.\mathbf{f}(\mathbf{u}) = 0, \tag{3.5.1}$$

where $\mathbf{u} \in \mathbf{R}^m$, $\mathbf{x} \in \mathbf{R}^n$, $t \geq 0$ and $\mathbf{f} : \mathbf{R}^m \to \mathbf{R}^m \otimes \mathbf{R}^n$ is some smooth mapping, assumed known. Comparing (3.5.1) with our balance law (1.2.2) in section 1.2, we see that $\mathbf{f}$ plays the role of flux density. Our aim is to discuss the relationship between (3.5.1) and the equation

$$\mathbf{u}_t + \nabla.\mathbf{f}(\mathbf{u}) = \varepsilon D \Delta \mathbf{u}, \tag{3.5.2}$$

in the limit $\varepsilon \to 0$ (D is a positive diagonal matrix), which may be thought of as (3.5.1) with diffusion, or with **viscosity**. In section 1.2, we saw how a diffusion term may arise by considering particles taking random walks, biased by a convective velocity term. In fluids, diffusion terms arise via the consideration of viscous forces, (see Box K or [37],[74]). (The Euler equations become the Navier-Stokes equations when considering conservation of momentum including the viscous terms.)

Either way, if (3.5.2) is our starting point, it is natural to try to exploit the simplification (3.5.1) wherever possible. Alternatively, if we seek to solve systems of the form (3.5.1), we shall show how (3.5.2) is useful (even though the viscous terms can only be perceived as artificial). We begin with a brief discussion of equations in the form of (3.5.1) and their weak solutions. To simplify things, we shall keep $n = m = 1$, the generalization being immediate for the most part (see [70]).

Consider the equation

$$u_t + f(u)_x = 0. \tag{3.5.3}$$

A classical solution is one for which u_t and $f(u)_x$ exist, are continuous, and satisfy (3.5.3). The point about such equations is that even for smooth initial data, prescribed at $t = 0$, say, the solution surface, $u(x,t)$, may fold up and develop an infinite derivative in finite time. When this happens, a discontinuity, or **shock**, will develop, and will itself be propagated according to the behaviour of the smooth solutions either side of it.

Consider (3.5.3) along with the initial data, parameterised by $s \in \mathbf{R}$:

$$\begin{aligned} x =& s, \\ u =& u_0(s), \\ t =& 0. \end{aligned} \tag{3.5.4}$$

Let us assume that $u(x,t)$ is a solution of (3.5.3-4). Now, along the curves satisfying

$$\frac{dx}{dt} = f_u(u), \tag{3.5.5}$$

we have u =constant. This is immediate from (3.5.3). Curves satisfying (3.5.5) are called **characteristics**. Now, we may solve

$$\begin{aligned} u =& u_0(s), \\ \frac{dx}{dt} =& f_u(u_0(s)), \quad t \geq 0, \\ x =& s \quad \text{when } t = 0 \end{aligned} \tag{3.5.6}$$

for the characteristics within our solution surface. Alternatively, if we are given $u_0(s)$, (3.5.6) implies that

$$\begin{aligned} u(x,t) =& u_0(s), \\ \text{where } x =& t f_u(u_0(s)) + s, \end{aligned} \tag{3.5.7}$$

is a solution of (3.5.3-4), provided that u_t and u_x remain finite. This method of solution is known as the **method of characteristics** (see [22]). The solution is valid provided that (3.5.7) can be inverted to write $s = s(x,t)$. (Here, we are of course assuming that $u_0(s)$ and f_u are given, continuous mappings.) This is trivial for $t = 0$, but if, for some $t > 0$, and some s,

$$t\frac{d}{ds}(f_u(u_0(s))) + 1 = 0,$$

then

$$u_x(x,t) = \frac{du_0}{ds}\frac{ds}{dx}$$

has become infinite and the surface may fold up, forcing u to become multivalued.

For example, suppose we take

$$f = \frac{u^2}{2}, \quad u_0(s) = \text{sech } s, \quad s \in \mathbf{R}.$$

We have $u(x,t) = \text{sech } s$, where

$$x = t \text{ sech } s + s$$

which is a differentiable surface up as far as

$$t = \min_{s \geq 0} \frac{\cosh^2 s}{\sinh s} = 2,$$

where the solution develops an infinite derivative before folding over. The solution is depicted schematically in Figure 3.17 for $t = 0, 1, 2, 3$.

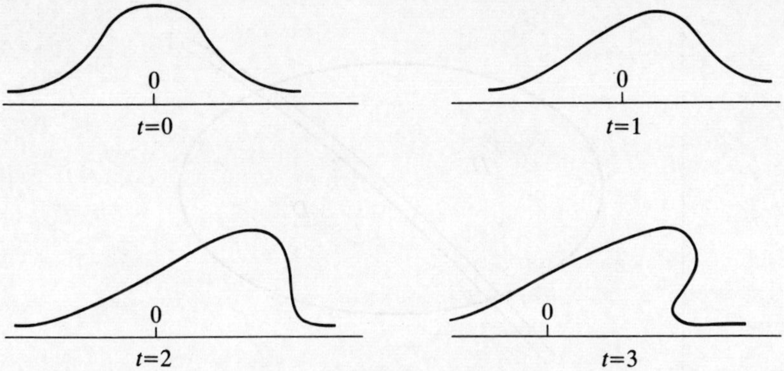

Figure 3.17: Development of a multivalued solution $u(x, t)$.

It is worth pointing out that classical solutions are unique while they exist (see [22]). However, in order to continue the solutions through the development of discontinuities, we must relax our notion of ***solutions*** to include discontinuous surfaces, $u(x, t)$. We do this as follows:

$u(x, t)$ is a **weak solution** of (3.5.3) in a domain $D \subseteq (x, t)$-space, provided

$$\int_D \phi_t u + \phi_x f(u)\, dx dt = 0 \tag{3.5.8}$$

for all C^1 functions, $\phi(x, t)$, defined on $\overline{D}$ (the closure of D), satisfying $\phi \equiv 0$ on ∂D.

Integration by parts immediately implies that classical solutions of (3.5.3) in D are also weak solutions. But (3.5.8) admits functions, $u(x, t)$, which are not even continuous. The key point is that (3.5.8) must hold for **all** admissible functions, ϕ.

Since we suspect from our example that we may wish to analyse solutions which possess finite discontinuities, it is revealing to see what (3.5.8) implies in such situations.

Suppose that u is a classical solution of (3.5.3) everywhere in D except along a curve, Γ, where it possesses a finite discontinuity or **shock**. Suppose Γ divides D into two parts, D_+ to the right, and D_- to the left (see Figure 3.18).

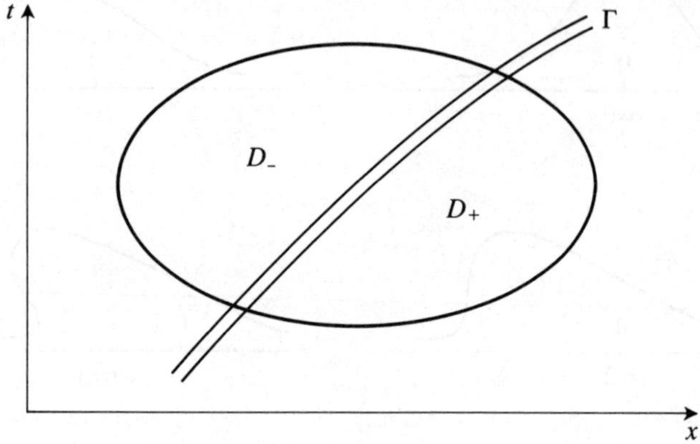

Figure 3.18: The shock curve, Γ.

Along Γ, let $u_\pm$ denote the limiting values of u as Γ is approached from with $D_\pm$. From (3.5.8),

$$\begin{aligned} 0 &= \int_D \phi_t u + \phi_x f(u)\,dxdt \\ &= \int_{D_+} \phi_t u + \phi_x f(u)\,dxdt + \int_{D_-} \phi_t u + \phi_x f(u)\,dxdt \\ &= \int_{D_+} (\phi u)_t + (\phi f(u))_x\,dxdt + \int_{D_-} (\phi u)_t + (\phi f(u))_x\,dxdt, \end{aligned}$$

since u is a classical solution in each of $D_\pm$. Now applying the divergence theorem to each integral, and using $\phi \equiv 0$ on ∂D, we obtain

$$0 = \int_{\Gamma\cap\partial D_+} \phi u\,dx - \phi f(u)\,dt + \int_{\Gamma\cap\partial D_-} -\phi u\,dx + \phi f(u)\,dt$$

where we integrate along Γ in the direction shown in Figure 3.18.

Thus,

$$0 = \int_{\Gamma\cap D} \phi([u]\,dx - \phi[f(u)]\,dt)$$

where $[u] = u_+ - u_-$ and $[f(u)] = f(u_+) - f(u_-)$. But ϕ was arbitrary in D, so we conclude that if $x = s(t)$ along Γ, then

$$[u]\frac{ds}{dt} = [f(u)]. \qquad (3.5.9)$$

Here, s_t is the **shock speed** (the speed at which the discontinuity moves relative to fixed x, in time), and (3.5.9) is known as the **jump** condition. If we set $z = x - s(t)$ so that the z-frame moves with the shock, (3.5.3) becomes

$$u_t + (f(u) - su)_z = 0,$$

so we see that (3.5.9) implies the continuity of flux across the shock (i.e. no sources or sinks occur at the discontinuity). Thus, (3.5.9) appears to be a natural condition that should be satisfied across all shocks.

Unfortunately, (3.5.9) is not enough to ensure that solutions (including those with shocks) are unique, in general. A further condition, known as an **entropy** condition must also be applied to rule out some pathological solution types, [70]. The origin of such a condition is not our concern here; we have seen enough of hyperbolic systems to be aware of some of the problems arising in their solution.

For our present purposes, we wish to compare the solution of (3.5.3) with its *viscous* counterpart

$$u_t + f(u)_x = \varepsilon u_{xx}. \tag{3.5.10}$$

Here, $\varepsilon > 0$ is small, so the diffusion term ought to be negligible provided that u_{xx} remains bounded. However, this can never be so (even when $\varepsilon \to 0$) in a neighbourhood of a shock in the corresponding solution of (3.5.3).

However, we may utilize the solution of (3.5.3) as an outer solution and locate a transition layer at any subsequent shock discontinuities.

Suppose the solution, u, of (3.5.3) has a shock at $x = s(t)$. Then we set $z = (x - s(t))/\varepsilon$ and write $u = U(z,t)$ in a neighbourhood of the shock. As $|z \to \pm\infty|$, we must have

$$U \to u_{\pm},$$

the values of u either side of the shock. Substituting for u in (3.5.10), we obtain

$$U_t = \frac{1}{\varepsilon}(U_{zz} + s_t U_z - f(U)_z).$$

Thus, up to $O(\varepsilon)$, we must solve

$$\begin{aligned} U_{zz} + s_t(t)U_z - f(U)_z &= 0 \\ U \to u_+(t), \quad z &\to \infty, \\ U \to u_-(t), \quad z &\to -\infty, \end{aligned} \tag{3.5.11}$$

for the transition profile. But (3.5.11) implies

$$U_z + s_t U - f(U) \equiv \text{constant}. \tag{3.5.12}$$

Now letting $z \to \pm\infty$ in (3.5.12), we see that

$$s_t u_+ - f(u_+) = s_t u_- - f(u_-)$$

which implies

$$s_t[u] = [f(u)]$$

which agrees precisely with (3.5.9). Thus, our shock condition falls out of the transition layer analysis by virtue of a first integral.

Moreover, it can be shown that for simple systems, solutions for which (3.5.11) yields transition profiles which converge, in the limit $\varepsilon \to 0$, to shocks satisfying the additional entropy condition given in [70]. (Notice that no transition layer profile may satisfy (3.5.11) if

$$s_t U - f(U) - (s_t u_\pm - f(u_\pm))$$

has any interior roots between u_+ and u_-.)

3.6 Piecewise linear systems

Consider the following system, which is a piecewise linear version of the FitzHugh-Nagumo system, (3.4.1):

$$\begin{aligned} u_t &= u_{xx} + f(u) - v, \\ v_t &= \varepsilon u, \end{aligned} \tag{3.6.1}$$

where $\varepsilon > 0$ and $f(u) = -u + H(u - a)$, $0 < a < 1$, and H is the usual Heaviside step function ($H(s) = 1$ for $s > 0$, $H(s) = 0$ for $s < 0$). Let us seek a travelling wave solution, $u = U(z)$, $v = V(z)$, where u is in the form depicted in Figure 3.19, $z = x - ct$ as usual, and c and $z_1 > 0$ are to be determined along with U and V.

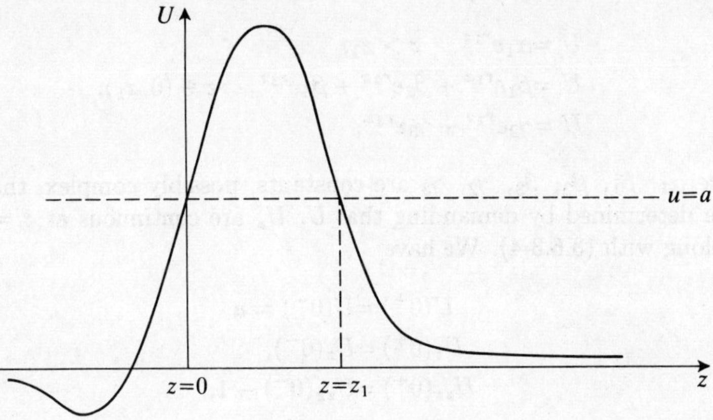

Figure 3.19: The pulse solution $U(z)$ for (3.6.1).

We have

$$\begin{aligned} U_{zz} + cU_z + f(U) - V &= 0, \\ cV_z + U\varepsilon &= 0. \end{aligned} \tag{3.6.2}$$

At $z = 0$, U increases through the switch state, $u = a$, and the nonlinearity jumps by 1, so there must be a commensurate jump of -1 in U_{zz} at $z = 0$, that is,

$$\left[U_{zz}\right]_{0^-}^{0^+} = -1. \tag{3.6.3}$$

Similarly, at $z = z_1$, U is assumed to decrease through a, so there is again a discontinuity in U_{zz}:

$$\left[U_{zz}\right]_{z_1^-}^{z_1^+} = 1. \tag{3.6.4}$$

Differentiating the first of (3.6.2) (away from $z = 0$ or z_1), we may eliminate $V(z)$ so that $U(z)$ satisfies

$$U_{zzz} + cU_{zz} - U_z + \frac{\varepsilon}{c}U = 0 \tag{3.6.5}$$

for $z \in \mathbf{R}\backslash\{0, z_1\}$.

The characteristic polynomial associated with (3.6.5) is

$$\rho(r) = r^3 + cr^2 - r + \frac{\varepsilon}{c}.$$

Since $\rho(0) > 0$, at least one root, say r_1, is negative and real. The other roots, r_2, r_3, must have positive real part since the coefficient of r in ρ is negative.

Thus, $U(z)$ is given by

$$\begin{aligned}U =&\alpha_1 e^{r_1 z}, \quad z > z_1,\\ U =&\beta_1 e^{r_1 z} + \beta_2 e^{r_2 z} + \beta_3 e^{r_3 z}, \quad z \in (0, z_1),\\ U =&\gamma_2 e^{r_2 z} + \gamma_3 e^{r_3 z},\end{aligned}$$

where α_1, β_1, β_2, β_3, γ_2, γ_3 are constants, possibly complex, that have to be determined by demanding that U, U_z are continuous at $z = 0$ and z_1, along with (3.6.3-4). We have

$$\begin{aligned}U(0^+) =&U(0^-) = a,\\ U_z(0^+) =&U_z(0^-),\\ U_{zz}(0^+) =&U_{zz}(0^-) - 1,\\ U(z_1^+) =&U(z_1^-) = a,\\ U_z(z_1^+) =&U_z(z_1^-),\\ U_{zz}(z_1^+) =&U_{zz}(z_1^-) + 1.\end{aligned}$$

For example, at $z = 0$, we have

$$\begin{aligned}\beta_1 + (\beta_2 - \gamma_2) + (\beta_3 - \gamma_3) =&0,\\ r_1\beta_1 + r_2(\beta_2 - \gamma_2) + r_3(\beta_3 - \gamma_3) =&0,\\ r_1^2\beta_1 + r_2^2(\beta_2 - \gamma_2) + r_3^2(\beta_3 - \gamma_3) =& - 1.\end{aligned} \tag{3.6.6}$$

Let $\rho_i = \frac{d\rho(r)}{dr}|_{r=r_i}$. Then the solution for the linear system

$$\begin{bmatrix}1 & 1 & 1\\ r_1 & r_2 & r_3\\ r_1^2 & r_2^2 & r_3^2\end{bmatrix}\begin{bmatrix}y_1\\ y_2\\ y_3\end{bmatrix} = \begin{bmatrix}0\\ 0\\ 1\end{bmatrix}$$

is simply $y_i = 1/\rho_i$, for $i = 1, 2, 3$. This elegant piece of algebra allows us to solve (3.6.6) and a similar set of conditions arising at $z = z_1$. We have

$$\begin{aligned}\beta_1 =& - 1/\rho_1,\\ \beta_2 - \gamma_2 =& - 1/\rho_2,\\ \beta_3 - \gamma_3 =& - 1/\rho_1,\\ e^{r_1 z_1}(\alpha_1 - \beta_1) =&1/\rho_1,\\ -e^{r_2 z_1}\beta_2 =&1/\rho_2,\\ -e^{r_3 z}\beta_3 =&1/\rho_3,\\ a =&\beta_1 + \beta_2 + \beta_3,\\ a =&\alpha_1 e^{r_1 z_1}.\end{aligned}$$

Together, these yield

$$e^{r_1 z_1} = (1 - \rho_1 a) \tag{3.6.7}$$

and

$$a = -\frac{1}{\rho_1} - \frac{1}{\rho_2}(1 - \rho_1 a)^{-r_2/r_1} - \frac{1}{\rho_3}(1 - \rho_1 a)^{-r_3/r_1}. \tag{3.6.8}$$

Now for a and ε fixed, the roots r_i and derivatives ρ_i (of p at $r = r_i$) may be thought of as functions of c. Hence, the *transcendental* equation (3.6.8) must be solved for the wave speed, c, and then all the constants $z_1, \alpha_1, \beta_1, \beta_2, \beta_3, \gamma_2, \gamma_3$ can subsequently be found using (3.6.7) and the equalities preceding it.

However, a more practical approach is to reverse the process and assume that c and ε are fixed (nonnegative). We may immediately determine the r_i and ρ_i to seek to solve (3.6.8) for the switch state a. Setting $s = (1 - \rho_1 a)$, (3.6.8) may be rewritten as

$$F(s) = 0$$

where

$$F(s) = 2 - s + s^{-r_2/r_1}\frac{\rho_1}{\rho_2} + \frac{\rho_1}{\rho_3}s^{-r_3/r_1}.$$

Since $s = (1 - \rho_1 a) = e^{r_1 z_1}$ and $z_1 r_1 < 0$, by definition, we must find a root of $F(s) = 0$ in (0,1). Using the relationships between the r_i and ρ_i, it is a straightforward matter to show

$$F(0) = 2, \quad F(1) = 0$$

$$\frac{dF(1)}{ds} = 0$$

$$\frac{d^2F}{ds^2}(1) = \frac{\rho_1}{r_1^2} - 2 = 1 + \frac{2c}{r_1} - \frac{1}{r_1^2}$$

Now, if F has a maximum at 1, then $F(s)$ must have a root in (0,1). This will be so provided

$$0 < r_1^3 + 2cr_1^2 - r_1,$$

but $r_1^3 + cr_1^2 - r_1 + \varepsilon/c = 0$, so this last condition holds provided

$$c < \sqrt{\frac{\varepsilon}{1 + 2\sqrt{\varepsilon}}}.$$

For fixed $\varepsilon > 0$, we can allow c to range over $\left(0, \sqrt{\varepsilon/(1 + 2\sqrt{\varepsilon})}\right)$, and solve $F = 0$ for $s \in (0, 1)$ (and hence a) numerically. We obtain a plot of wave-speed, c, versus switch state, a, at which pulse solutions exist. In Figure

3.20, we have drawn a schematic version of such a plot (the reader may either carry out the necessary computations, if desired, or else consult [48] or [67] for an account of this result).

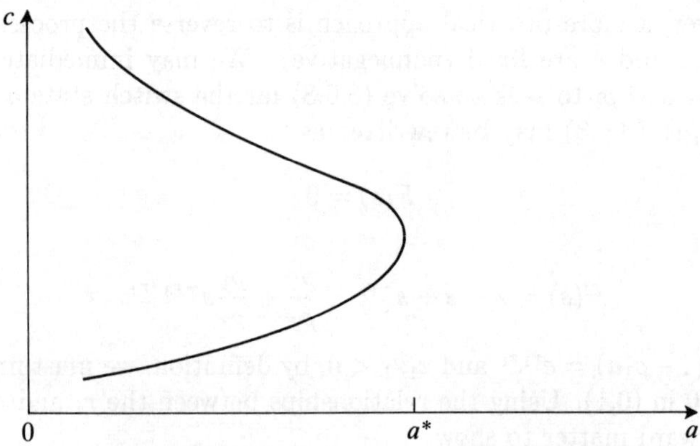

Figure 3.20: c versus a for pulse solutions for (3.6.1).

The important point to note from Figure 3.20 is that it turns out that, for a less than some constant a^*, (3.6.1) possesses two pulse solutions, the faster one corresponding to the transition layer type solution that may be obtained by the methods of section 3.4.

As in section 3.4, we may also utilize this technique to seek travelling periodic solutions.

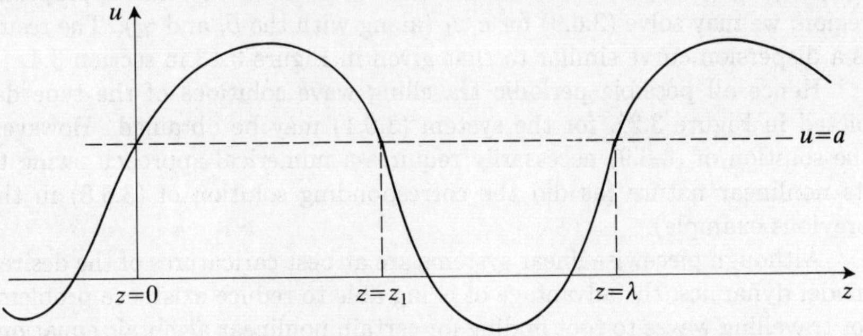

Figure 3.21: A periodic wave with wavelength λ.

Suppose we seek a solution of the form depicted in Figure 3.21, where the wavelength, $\lambda > 0$, and $z_1 \in (0, \lambda)$ are to be determined along with the wave speed, c, and solution, U,

$$\begin{aligned} U =& \beta_1 e^{r_1 z} + \beta_2 e^{r_2 z} + \beta_3 e^{r_3 z} \qquad z \in [0, z_1] \\ U =& \gamma_1 e^{r_1 z} + \gamma_2 e^{r_2 z} + \gamma_3 e^{r_3 z} \qquad z \in [z_1, \lambda] \end{aligned}$$

which satisfies the periodic, continuity, and jump conditions:

$$\begin{aligned} U(0^+) =& U(\lambda^-) = a \\ U_z(0^+) =& U_z(\lambda^-) \\ U_{zz}(0^+) =& U_{zz}(\lambda^-) - 1 \\ U(z_1^+) =& U(z_1^-) = a \\ U_z(z_1^+) =& U(z_1^-) \\ U_{zz}(z_1^+) =& U(z_1^-) + 1. \end{aligned}$$

Using algebra similar to that of the travelling pulse case, we obtain

$$\begin{aligned} (\beta_i - \gamma_i) e^{\lambda r_i} &= -1/\rho_i \qquad i = 1, 2, 3, \\ (\beta_i - \gamma_i) e^{z_1 r_i} &= -1/\rho_i \qquad i = 1, 2, 3, \\ \sum_{i=1}^{3} \beta_i &= \sum_{i=1}^{3} \beta_i e^{r_i z_1} = a. \end{aligned} \tag{3.6.9}$$

For ε and a fixed, (3.6.9) yields eight conditions for nine unknowns (c, $\lambda, z_1, \beta_i, \gamma_i$). Hence if the wavelength, λ, is varied in some appropriate region, we may solve (3.6.9) for c, z_1 (along with the β_i and γ_i). The result is a dispersion curve similar to that given in Figure 3.13 in section 3.4.

Hence all possible periodic travelling wave solutions of the type depicted in Figure 3.21, for the system (3.6.1) may be obtained. However, the solution of (3.6.9) necessarily requires a numerical approach owing to its nonlinear nature (as did the corresponding solution of (3.6.8) in the previous example).

Although piecewise linear systems are at best caricatures of the desired model dynamics, the advantage of being able to reduce existence problems for travelling waves to root finding for certain nonlinear algebraic equations is a strong one.

For (3.6.1), we were able to obtain a full description of the travelling pulse and periodic wave phenomena, albeit modulo some numerical computation at the last stage. This enhances our intuition concerning such waves when we come to analyse the full, nonlinear systems via transition layer analysis or other means.

Piecewise linear dynamics may be used for a variety of nonlinear systems, and have been employed in a number of situations [30], [67]. The relatively unsophisticated mathematics required in such problems is attractive, especially when one is making a preliminary approach to a new model or new area of applications.

Exercise 3.1

Consider the equation

$$u_{zz} - cu_z + e^u \sin u = 0.$$

For $u \in (0, 2\pi)$, the nonlinear term is of the same *cubic* form as that discussed earlier. Let $(c_0, \phi_0(z))$ denote the unique value of c and the corresponding solution satisfying

$$u \to 0 \text{ as } z \to \infty$$
$$u \to 2\pi \text{ as } z \to -\infty.$$

For $p = 1, 2, \ldots$, define

$$c_n = e^{\pi} c_{p-1}$$
$$\phi_n^{(z)} = \phi_{p-1}(ze^{\pi}) + 2\pi.$$

Show that (c_p, ϕ_p) are homoclinic solutions of the above equation satisfying

$$\phi_p \to 2\pi p \text{ as } z \to \infty$$
$$\phi_p \to 2\pi(p+1) \text{ as } z \to -\infty.$$

By considering the trajectories in the phase plane, as c varies, deduce that there exists a unique value of $c \in [c_0, c_1]$ such that there is a solution $\psi(z)$ satisfying

$$\psi \to 0 \text{ as } z \to \infty$$
$$\psi \to 4\pi \text{ as } z \to -\infty.$$

How far can this result be generalized? Can you show that there exists a unique value of c such that there is a solution satisfying

$$\psi \to 2\pi p \text{ as } z \to \infty$$
$$\psi \to 2\pi q \text{ as } z \to -\infty$$

for any integers, $q > p \geq 0$?

Exercise 3.2

Consider the dispersal of a chemical tracer through a uniform, one-dimensional, saturated porous medium, which may leave solution and sorb to the rock volume:

$$\phi c_t = Dc_{xx} - Vc_x - s_t.$$

Here ϕ is the porosity (the volume fraction available to the ground water), D is the intrinsic diffusion, and V is the Darcy velocity (the equivalent velocity of the water through empty space: V/ϕ = the average water velocity in the porespace), and s denotes the amount of sorbed tracer per unit volume.

Let p denote the density of sorption sites, scaled so that each site has capacity for a unit amount of the sorbed tracer. If no tracer is present then $p \equiv p_0 > 0$, constant, everywhere.

Assume that sorption takes place via the (reversible) interaction

$$[p] + [c] \rightleftharpoons [s],$$

which equilibrates on a fast time-scale compared to that associated with tracer dispersal.

We have

$$s_t = \frac{\phi cp}{\varepsilon} - \frac{sk}{\varepsilon}$$
$$p_t = -s_t$$

where $k\varepsilon^{-1}$ and ε^{-1} are very large rate constants. Let us assume $\varepsilon << 1$. Clearly

$$s + p = p_0.$$

Show that sorption-desorption equilibrium (valid on all but short time-scales, $O(\varepsilon, \varepsilon/k)$) yields

$$s = \frac{\phi c p_0}{(k + \phi c)}.$$

This relationship between the dissolved concentration c and the sorbed density s is known as a **Langmuir isotherm**. We shall assume it is valid identically on the time-scale associated with the tracer transport problem. Substituting this into the transport equation, we obtain

$$\left(\phi c + \frac{\phi c p_0}{(k + \phi c)}\right)_t = Dc_{xx} - Vc_x.$$

Consider the situation where a tracer is supplied at a known concentration, c_0, at a far distance ($x \to -\infty$), and is dispersed throughout previously unoccupied water-filled rock.

We have the boundary conditions

$$c \to c_0 \text{ as } x \to -\infty$$

$$c \to 0 \text{ as } x \to +\infty.$$

Let us seek a travelling wave solution

$$c(x, t) = C(z),$$

where $z = x - \theta t$ is a moving frame, with wave speed, θ. Notice that if $\theta > 0$ the wave motion is to the right (similarly $V > 0$ represents a rightward water flow).

We obtain

$$\left(VC - \theta\left(\phi C + \frac{\phi C p_0}{(k + \phi C)}\right)\right)_z = DC_{zz}.$$

Integrating once, using $C \to 0$ as $z \to +\infty$, show that

$$DC_z = VC - \theta\left(\phi C + \frac{\phi C p_0}{(k + \phi C)}\right).$$

Now C must be nonnegative, and satisfy

$$C(-\infty) = c_0, \text{ and } C(+\infty) = 0.$$

Show that this is possible if and only if

$$\theta = \frac{V}{\phi} \frac{(k + \phi c_0)}{(k + \phi c_0 + p_0)},$$

and the quantity

$$VC - \theta\Big(\phi C + \frac{\phi C p_0}{(k + \phi C)}\Big)$$

is strictly negative in $(0, c_0)$ (note that this vanishes at both 0 and c_0).

Show that this last condition will hold so long as $V > 0$. Hence deduce that, for $V > 0$, C may be obtained from:

$$\int_{c_0/2}^{C} \frac{D\, d\tilde{c}}{V\tilde{c} - \theta\Big(\phi\tilde{c} + \frac{\phi\tilde{c}p_0}{(k+\phi\tilde{c})}\Big)} = z + \text{constant},$$

for $C \in (0, c_0)$, where θ is given above.

The integration may be carried out analytically ...

What is the effect upon the wave profile of changing D?

Exercise 3.3

Consider the problem (3.4.12):

$$0 = \chi_{\zeta\zeta} + \gamma\chi_\zeta + f(\chi) - \alpha,$$

$$\chi \to h_-(\alpha) \text{ as } \zeta \to \infty,$$

$$\chi \to h_+(\alpha) \text{ as } \zeta \to -\infty;$$

see Figure 3.6 for an explanation of the pseudo inverses $h_\pm$.

Here, α is fixed within the range $(V_{\min}, V_{\max})$ so that f has three roots, $h_-(\alpha)$, $h_+(\alpha)$, and one in between.

Solve this explicitly for χ and γ in the case where f is the cubic $u(u-a)(1-u)$. (Compare this with (1.5.3) where a unique solution was obtained.)

Show that the relationship between α and γ is of the form shown in Figure 3.8.

Exercise 3.4

Consider a piecewise linear version of the problem posed in the previous question.

Define the function

$$g(\chi) = -\chi + H(\chi - a)$$

where a is a constant in (0,1), and H is the Heaviside step function ($\equiv 1$ at all positive arguments, zero otherwise). Define the pseudo inverses

$$h_-(y) = -y, \quad y > -a,$$

$$h_+(y) = -y, \quad y < 1 - a,$$

so that for $y \in (-a, 1-a)$ we have

$$y = g(h_-(y)) = g(h_+(y)).$$

Then for each α fixed in $(-a, 1-a)$ solve the problem

$$0 = \chi_{\zeta\zeta} + \gamma\chi_\zeta + g(\chi) - \alpha,$$

$$\chi \to h_-(\alpha) \text{ as } \zeta \to \infty,$$
$$\chi \to h_+(\alpha) \text{ as } \zeta \to -\infty,$$

for $\chi(\zeta)$ and γ, a real constant (by piecing together solutions of the linear systems).

Sketch the relationship between α and γ in this case. How does it differ from the form shown in Figure 3.8, deduced in the previous question?

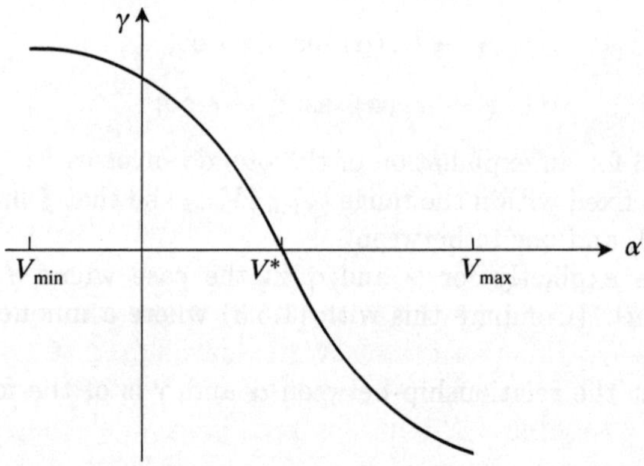

4 A geometrical theory for waves

4.1 Introduction

In Chapter 3, we were concerned with plane wave solutions of reaction-diffusion systems. Even for small systems, the existence and stability problems for these travelling waves are challenging. In the present chapter, we shall take a further step towards a more complete understanding of wave-like phenomena in reaction-diffusion. More precisely, we shall consider waves which propagate through $\mathbf{R}^3$, or over surfaces, which do not necessarily have planar symmetry.

A plane wave front moving through $\mathbf{R}^3$ has constant velocity, in the forward, normal direction, at all points. If the plane geometry of the wave is distorted, we expect that there will be an effect on the normal velocity, which will now vary locally across the wave front.

In order to begin our discussion, we must have a general idea as to what constitutes a wave or wave-like structure. We shall use these terms to refer to solutions which cause the state of the system to be shifted through some qualitative sequence, successively in adjacent neighbourhoods within the domain. For example, a wave front changes the state of the system from one steady state to another as it passes over. By choosing an intermediate state, we can home in on the surface of points in the domain at which this distinguished state is achieved instantaneously. As time evolves, the progress of our surface is representative of the wave-front. The normal velocity and curvature of the surface will vary locally as well as with time. Of course, had we chosen an alternative intermediate (distinguished) state, we would have obtained a different surface, probably propagating according to slightly different rules. A full knowledge of the behaviour of each and every such surface would require a complete knowledge of the solution to the original reaction-diffusion problem. So how can these notions help us? Many initial value problems, on large domains, can yield solutions with local nonplanar wave-like structure, so are we to attempt to resolve all such problems at once?

The answer lies in restricting the nature of the systems that are under consideration. In fact, we shall home in, for the most part, upon those systems possessing a *fast* state variable which dominates the behaviour of the system. Moreover, we shall also assume that the wave solutions possess transition-layers in which the dominant state variable is shifted rapidly from one regime to another. This last simplification has two major advantages.

Firstly, it suggests that we try to solve the problems by a transition-layer approach (which proved successful for a variety of problems in Chapter 3), and that it is the behaviour of the solution within the transition-layers that is important in describing the spreading of the waves.

Secondly, since the state is changing rapidly across the transition-layer, the surfaces at which the solution achieves distinguished intermediate states are all very close together, and hence must exhibit a similar geometry and normal speed of propagation. Thus we hope to avoid the complexities involved in the general situation indicated earlier.

Assuming that the waves are quiescent away from the transition layers, the problem reduces to the following:
(i) to describe how the solution varies across the transition layer;
(ii) to describe how the geometry and motion of the transition layer evolve in time.

Here, we shall employ a slight dichotomy; although our transition layers are (slim) three dimensional slices through $\mathbf{R}^3$, by hypothesis, there is some small parameter, say ε, present in the system which controls the width of the layer. As $\varepsilon \to 0$, the layer must collapse towards a two dimensional surface. The limiting surface locates the transition-layer solution, so that it makes sense to think of this as a kind of fundamental surface which describes the wave. We shall obtain equations which control the geometry and motion of such surfaces. In any real example, we must always backtrack and locate the appropriate transitional profiles within a suitable neighbourhood of our surface, so as to obtain a continuous asymptotic solution.

Our discussion will involve special types of wave-like solutions such as spiral and scroll waves, which themselves evolve through a cycle until returning to a previous configuration. We shall also consider waves which evolve from initial, general, configurations, and try to make statements as to their long-term behaviour.

In section 4.3, where the main equations are derived, we must do some basic geometry. There, we shall use $\mathbf{r}$ to denote points in $\mathbf{R}^3$, rather than $\mathbf{x}$ (which was previously used to denote general points within our domains). This is just personal preference (the very sight of coordinate transformations in $\mathbf{R}^3$ is enough to induce me to adopt the use of $\mathbf{r}$!), and will not, I hope, cause any confusion.

4.2 Spirals and scrolls

The main motivation for studying nonplanar, wave-type solutions for reaction-diffusion equations, lies in their applicability to problems in physiology, biology, and chemistry. For example, the Hodgkin-Huxley equations [36] (which are a system of reaction-diffusion equations) provide a model for the electrical activity in membranes of living organisms. One of the state variables represents the potential difference across the membrane. This is subject to the electrical currents conducted tangentially to the membrane in both the external and internal ionic fluids, as well as trans-membrane currents due to the gating of specific charged ions through active channels as well as the capacitance of the membrane itself.

In one dimension, we may utilize the Hodgkin-Huxley equations as a model for the propagation of action potentials along an unmyelinated nerve axon (in fact, this was the original problem for which the equations were derived, [36]). The FitzHugh-Nagumo equations, (see Chapter 3), constitute a similar, but more qualitative, model.

Outside of one dimension, we may employ the Hodgkin-Huxley equations to model the electrical activity taking place in cells within the human heart. Immediately, our domain Ω possesses a subtle geometry, and it may be hoped that our solutions display both normal modes of operation and any possible malfunctions, such as cardiac arrhythmias. There is a vast literature concerning the physiology and electrical behaviour of cardiac tissue, and we cannot hope to do this justice here. However, in his book [77], A.T. Winfree supplies an absorbing account of these matters (and much more); in particular, Winfree is able to introduce many concepts concerning the geometry and motion of wave propagation in a way that is attractive to the mathematician and nonmathematician alike.

For our present purposes, it is enough to appreciate that waves of depolarization of the membrane potentials in cardiac tissue are the precursors of the coordinated muscle contractions which ensure that the heart can fulfil its normal function. Any small disturbance to the waves will, hopefully, have a negligible effect; but what of larger perturbations? If groups of cells become inactive or damaged, for example, this may result in the normal (cyclic) mode of operation becoming degenerate, possibly inducing rogue waves propagating in diverse directions, leading to complex arrythmias, distorting the coherence of the heart's essential cycle.

Obviously, an accurate model of such cardiac arrythmias must incorporate many externalities: the geometry of the tissue, the source, strength, and nature of the initial stimulus for each cycle, the nature of the abnormalities to be encountered. However, the mathematician is unable to hide behind such complicated prerequisites. The plain fact is that the phenomena associated with the normal and abnormal modes of wave propagation in such problems are much more prevalent than we may at first think.

In Box J, we give a brief introduction to a, by now, famous chemical reaction, the Belousov-Zhabotinsky reaction. Consisting of a solution of unstirred chemical components, there can be no doubt that the process is correctly described by a system of reaction-diffusion equations.

Of course, we must first be in a position to write down the dynamics of the reaction terms, with some justification, but thereafter, we should, in theory, be able to replicate any real experiment with real solutions to the model.

By adjusting the recipe, the Belousov-Zhabotinsky reaction can be put into an excitable mode so that the experiments display the same qualitative behaviour as the electrical membranes discussed earlier. (Recall that a system is **excitable** if, given a sufficient stimulus or initial perturbation, it exhibits a large excursion in the state space before returning to its previous quiescent stable regime.) The similarity between the excitable membranes and the Belousov-Zhabotinsky reagent goes further. In [77], Winfree states that:

They are so close in form and in qualitative behavior that the reagent might be fairly called aqueous solution of the equations of electrophysiology (pun intended).

Thus, in the Belousov-Zhabotinsky reagent, we have a model for wave propagation within excitable media. Any mathematical theory must first be able to explain the behaviour exhibited by the reagent; but this is a demanding standard to be set.

When the Belousov-Zhabotinsky reagent is placed in thin films, in dishes, or absorbed in filter paper, a whole world of spiral wave patterns becomes apparent (see Figure 4.1). The spiral waves themselves provide a mathematical challenge: how can we set about such solutions for reaction-diffusion systems? The state of the reaction fluctuates widely as we move through the propagating wave, so the system must be far from equilibrium; and what of the core of the spiral? A rigorous analysis of such phenomena is still awaited, but the subject matter of this chapter (developed initially [39] to examine spiral waves in the plane) is able to provide some clues.

Outside of two dimensions, there is worse to come. In three dimensional domains (say, tightly stoppered test-tubes), Welsh et. al. [76] have documented more esoteric phenomena. Scroll waves, toroidal scroll waves, and linked-twisted scroll waves are all apparent to some degree. Clearly, the simple spiral is the tip of an exciting iceberg of wave-like structure.

Over a number of years, A.T. Winfree [77] has pioneered the consideration of such phenomena. By setting aside the equations and their solutions, he was able to discuss the geometry and motion of propagating (often oscillatory) wave structures. By topological (continuity) arguments,

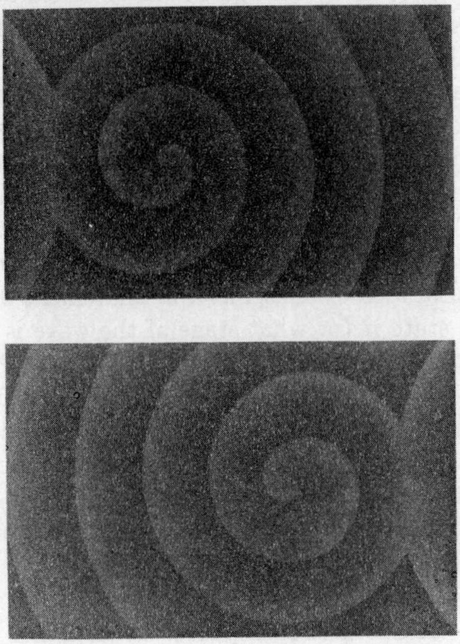

Figure 4.1: Spiral waves

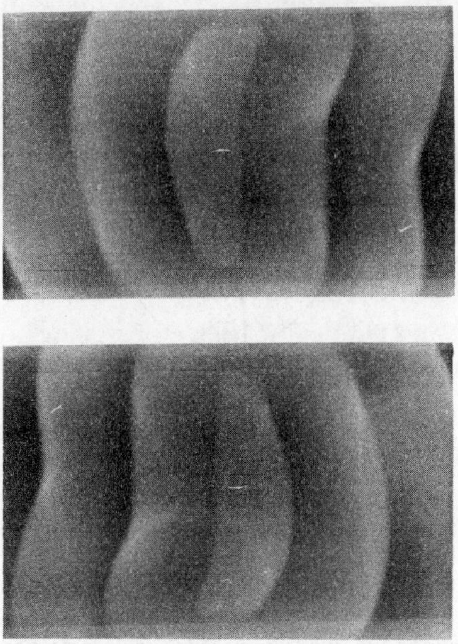

Figure 4.2: Toroidal scrolls (from [67], with permission).

it is possible to rule out many configurations of linked and twisted scrolls, but we are nevertheless no nearer to any analytical statements concerning solutions to excitable reaction-diffusion systems.

The basic ideas contained within this approach are as follows.

If we were to observe a wave passing by, we would see the state of the reaction being excited away from its quiescent stable regime, then moving through a number of distinct stages before returning to the original unexcited state. The next wave would then pass, repeating the same qualitative process. This circular evolution, associated with excitable systems, gives rise to a natural idea of phase, as a measurement of how far through the cycle the local state is (or what stage of the wave is currently passing through).

The **phase** of the system should be thought of as a functional, defined on the state space, taking values on the unit circle. We can then get a snapshot of a wave, at any time, by investigating points in real space, whose states are of the same phase. We obtain surfaces containing points of identical phase, and their evolution through the domain, in time, is representative of the original wave structure.

This is fine for simple wave forms, but for rotating wave phenomena, it is not sufficient. At the centre of spirals, for example, we have phaseless points – points where all phases converge as in Figure 4.3. (What time is it at the North Pole?)

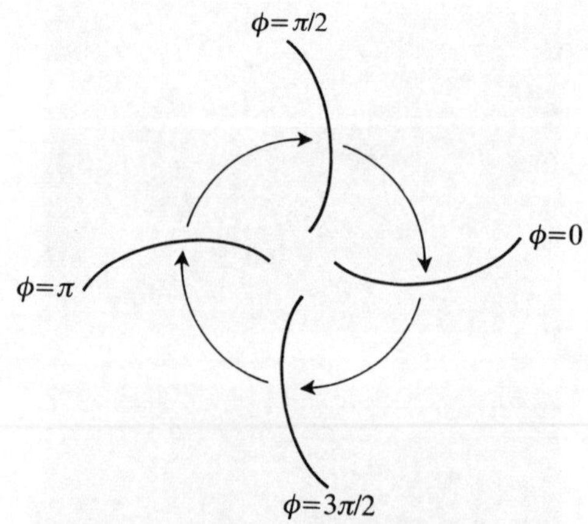

Figure 4.3: Curves of constant phase, ϕ, rotating about a phaseless core.

In three dimensions, we visualize the phaseless points as forming arcs or curves. Such curves are called **singular filaments** [78]. By fixing a critical phase, we can see how the corresponding constant phase surface propagates around the singular filament as time passes. For example, a simple linear scroll wave can be visualised as a linear filament and a scrolling surface (locating points of some distinguished phase) rotating about it, as in Figure 4.4.

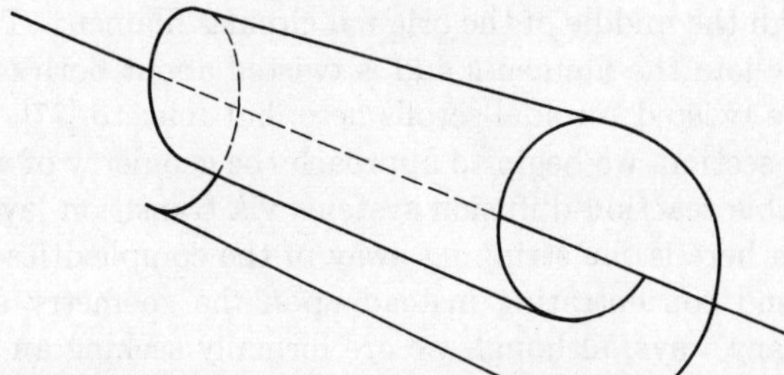

Figure 4.4: A linear scroll wave in $\mathbf{R}^3$.

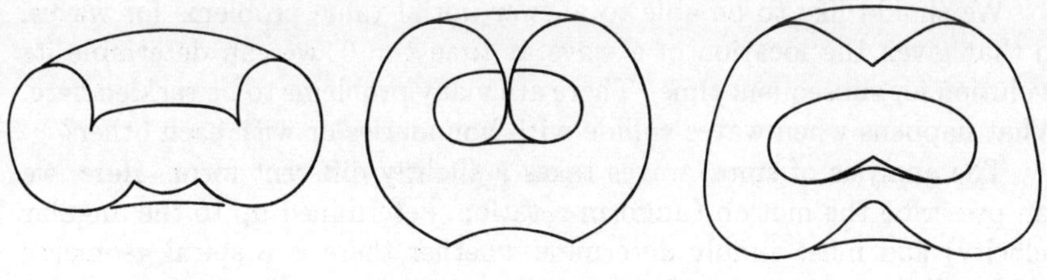

Figure 4.5: A toroidal scroll wave in $\mathbf{R}^3$.
Three stages within a single cycle.

If a linear scroll is bent around, in a circle, so that the filament and scrolling surface are joined, we obtain a toroidal scroll wave. Such a wave possesses a singular filament in the form of a ring, as well as a point somewhere in the middle where the scrolling surface converges, before dividing into two components (one equivalent to a sphere propagating off to ∞, and the other restarting the original cycle) see Figure 4.5.

If the linear scroll were twisted before being bent round and joined up, we would obtain a twisted toroidal scroll. It is a good exercise to try to sketch such a wave. A further complication also arises: the twisted surface can not be joined up smoothly unless we introduce a new singular filament threaded through the middle of the original circular filament. The scrolling surface can now join the filaments and is twisted about both of them. We shall not pursue twisted toroidal scrolls here, but refer to [77].

In the next section, we begin to approach the geometry of wave propagation in excitable reaction-diffusion systems via transition layer analysis. The central aim here is the stripping away of the complexities of the particular sytem and concentrating instead upon the geometry and motion of waves. In many ways, although we are formally seeking an asymptotic solution to a reaction-diffusion problem, this construction is analogous to Winfree's abstract wave construction. Moreover, the asymptotic approach presented here formally breaks down at the singular filaments introduced above (just as the phase approach does). However, we shall be able to gain some qualitative information about the way in which the waves propagate, which is itself of use in a variety of reaction-diffusion problems.

As a first step, we can consider the motion of waves moving through two- and three-dimensional domains. Here, we mean a sharp transition layer which sweeps the state variables from one regime to another as it propagates through space in time.

We should like to be able to answer initial value problems for waves, so that given the location of a wave at time $t = 0$, we can determine its evolution for subsequent time. There are many problems to be tackled here. What happens when waves collide with boundaries or with each other?

The analysis of spiral waves takes a slightly different form. Here, we can prescribe the motion (uniform rotation, determined up to the angular velocity) and must simply determine whether there is a spiral geometry which will entertain such behaviour. The toroidal scroll waves are not quite so straightforward, as we cannot make a guess as to the motion *a priori*, and must instead seek to solve for the geometry and motion simultaneously (as we would in the initial-value problems).

We shall return to spirals and scrolls later after our initial transition-layer considerations.

As for the cardiac arrhythmias, we should ideally like to have some means of simulation and explanation for such behaviour. It is possible that the next few years will produce a vast growth of interest in such problems, particularly as more analytical techniques are utilized in their discussion and resolution.

4.3 Geometrical theory

The aim of this section is to consider a simple model of reaction-diffusion wave propagation.

We shall show how the problem of three-dimensional wave propagation can be compared with that for wave propagation in one dimension by demanding that the front of the wave is located on a surface which propagates according to

$$N + \varepsilon K = c \tag{4.3.1}$$

(where N is normal velocity, K is twice the mean curvature, and ε and c are constants independently obtained from the original reaction-diffusion equation).

We shall refer to (4.3.1) as the **eikonal equation**.

The basic idea is to represent the wave front by a local transition layer exploiting a small scaling constant in the original reaction-diffusion system.

It is simplest initially to focus upon examples. So, although the theory may be applied to complicated coupled systems, it may serve us well to consider the equation

$$\varepsilon u_t = \varepsilon^2 \Delta u + u(1-u)(u-a), \tag{4.3.2}$$

for a constant $a \in (0,1)$.

The associated one-dimensional travelling equation (see section 1.5), for plane wave solutions $u = w(\xi)$, where $\xi = (x-ct)/\varepsilon$,

$$0 = w_{\xi\xi} + cw_\xi + w(1-w)(w-a), \tag{4.3.3}$$

has a solution, such that

$$w(\xi) \to 1 \text{ as } \xi \to -\infty,$$

$$w(\xi) \to 0 \text{ as } \xi \to +\infty,$$

provided $c = \sqrt{2}(1/2 - a)$. In fact $w(\xi) = (1 + \exp(\xi/\sqrt{2}))^{-1}$.

Now for any wave-type behaviour supported by (4.3.2), we can think of the solution, u, as being zero ahead of the wave and identically one behind the wave; and exploit $1 \gg \varepsilon > 0$ in order to obtain a transition layer (asymptotic) solution in a neighbourhood of the (nonplanar) wave front.

In other examples, we can still think of (4.3.2) as providing the wave-front behaviour and imagine slow recovery processes following along behind the front which return u to the unstimulated equilibrium $u = 0$.

We shall derive (4.3.1) in Box I by considering the transition layer of (4.3.2) at the wave front, and comparing the resulting ordinary differential equation with (4.3.3).

The key point is that the Laplacian term yields the curvature correction εK in the equation (4.3.1). If $\varepsilon = 0$ in (4.3.1), and thus $N = c$, we have the eikonal equation of geometrical optics.

However, even for small ε, the curvature correction may not be negligible (at the centre of spiral waves, for example [39]). Thus we cannot *a priori* ignore this term, and (4.3.1) should be regarded as a singular perturbation of the geometrical optics eikonal equation.

If a wave is in contact with a boundary (i.e. a boundary of the domain in which the original reaction-diffusion system is satisfied), we impose boundary conditions which are consistent with the imposition of Neumann boundary conditions upon the reaction-diffusion system. For a wave meeting the boundary, this simply requires that the edge of the wave-front must propagate across the boundary so that the wave always meets the boundary orthogonally (i.e. the normal vectors to the surface locating the wave front and the boundary surface are orthogonal).

Box I: The eikonal equation

Consider

$$\varepsilon u_t = \varepsilon^2 \Delta u + F(u), \tag{1}$$

where $u \equiv u_1$ and $u \equiv u_2$ are stable rest states for the associated spatially independent problem.

In addition, we shall suppose that (1) admits a one-dimensional travelling wave solution, $\phi(z)$, such that

$$0 = \phi'' + c\phi' + F(\phi), \quad z \in \mathbf{R} \tag{2}$$

and

$$\phi \to u_1 \text{ as } z \to +\infty,$$

$$\phi \to u_2 \text{ as } z \to -\infty,$$

for some wave speed c.

We construct an asymptotic solution of (1) as follows.

Suppose that there is some oriented surface M moving through $\mathbf{R}^3$ such that $u \approx u_1$ ahead of the surface (in the oriented normal direction), and $u \approx u_2$ behind the surface. At the surface, we must locate a transition layer so as to match the two constant outer solutions.

For each time t, fixed in some interval, let $\mathbf{r}^*(\eta, \lambda, t)$ range over a neighbourhood in M as real parameters η and λ vary over some suitable domain.

We may introduce the stretched normal coordinate ξ via the mapping

$$\mathbf{r} = \mathbf{r}^* + \varepsilon\xi\mathbf{n}(\mathbf{r}^*) + O(\varepsilon^2\xi^2), \tag{3}$$

so that for t fixed, (ξ, η, λ) is a local coordinate system.

Here, $\mathbf{n}(\mathbf{r}^*)$ is the oriented normal to M at $\mathbf{r}$: without loss of generality, we choose

$$\mathbf{n}(\mathbf{r}^*) = \frac{(\mathbf{r}^*_\eta \times \mathbf{r}^*_\lambda)}{|\mathbf{r}^*_\eta \times \mathbf{r}^*_\lambda|}.$$

The idea here is that for each ξ fixed in (3), η and λ parameterize a surface ($\xi = 0$ corresponds to M). We shall assume that (3) is defined in such a way that for each ξ-constant surface, $\mathbf{r}_\xi/\varepsilon$ always denotes the corresponding unit normal vector.

Now, we set $u = \psi(\xi)$ in a neighbourhood of M and demand that u satisfy (1) together with $\psi \to u_1$ as $\xi \to \infty$ and $\psi \to u_2$ as $\xi \to \infty$.

Firstly, since $d/dt\ \mathbf{r} = 0$ in (3), we have

$$\mathbf{r}_t + \mathbf{r}_\xi\xi_t + \mathbf{r}_\eta\eta_t + \mathbf{r}_\lambda\lambda_t = 0.$$

Thus, setting $\xi = 0$, we may obtain

$$\frac{\mathbf{r}_t.\mathbf{r}_\xi}{\varepsilon} = -\xi_t\varepsilon.$$

Notice, $\mathbf{r}_t.\mathbf{r}_\xi/\varepsilon = N$ is the normal velocity of the constant-ξ surface. Furthermore, since $u = \psi(\xi)$, we have

$$\varepsilon(\partial u/\partial t) = \varepsilon(d\psi/d\xi).\xi_t = -N(d\psi/d\xi).$$

Secondly, if we assume that ξ, η, and λ are orthogonal curvilinear coordinates in a neighbourhood some point $(\xi_0, \eta_0, \lambda_0)$, we have

$$\begin{aligned}
\Delta u =& \frac{1}{|\mathbf{r}_\xi||\mathbf{r}_\eta||\mathbf{r}_\lambda|}\left(\frac{|\mathbf{r}_\eta||\mathbf{r}_\lambda|}{|\mathbf{r}_\xi|}\psi_\xi\right)_\xi \\
=& \frac{\psi_{\xi\xi}}{|\mathbf{r}_\xi|^2} + \frac{\psi_\xi(|\mathbf{r}_\eta|.|\mathbf{r}_\lambda|)_\xi}{(|\mathbf{r}_\eta||\mathbf{r}_\lambda||\mathbf{r}_\xi|^2)} \\
=& \frac{\psi_{\xi\xi}}{\varepsilon^2} + \psi_\xi\left(-\frac{\mathbf{r}_{\eta\eta}.\mathbf{r}_\xi}{|\mathbf{r}_\eta|^2} - \frac{\mathbf{r}_\lambda.\mathbf{r}_\xi}{|\mathbf{r}_\lambda|^2}\right)\frac{1}{\varepsilon^2} \\
=& \frac{\psi_{\xi\xi}}{\varepsilon^2} + \frac{\psi_\xi K}{\varepsilon},
\end{aligned}$$

where

$$K = -\left(\frac{\mathbf{r}_{\eta\eta}}{|\mathbf{r}_\eta|^2} + \frac{\mathbf{r}_{\lambda\lambda}}{|\mathbf{r}_\lambda|^2}\right) . \frac{\mathbf{r}_\xi}{\varepsilon}$$

is twice the mean curvature of the surface $\xi =$ constant. If ξ, η and λ are not orthogonal, we can still obtain the same result, but K is given by a more complicated formula and the algebra is more involved [12]. In the applications that we consider, we always have ξ,η, and λ orthogonal.

Thus, if $u = \psi(\xi)$ satisfies (1), we have

$$\psi_{\xi\xi} + (N + \varepsilon K)\psi_\xi + F(\psi) = 0. \tag{4}$$

Now we note that, using (3), we can show $N = N|_{\xi=0} + O(\varepsilon^2)$; also it is clear that $K = K|_{\xi=0} + O(\varepsilon)$. Hence, ignoring terms $O(\varepsilon^2)$, $N + \varepsilon K$ is independent of ξ. Thus, we shall evaluate N and K at $\xi = 0$, so that they denote the normal velocity and twice the mean curvature of M respectively.

Now we compare (4) with (2) to see that $\psi = \phi(\xi)$, when $c = N + \varepsilon K$ is precisely the solution we seek for ψ, and hence our transitional inner solution for u at M.

The condition

$$c = N + \varepsilon K$$

is an equation for the time evolution of the surface M, which locates the wave. We refer to it as the **eikonal equation.**

Here, we have derived the eikonal equation in $\mathbf{R}^3$. In [39] and [25], there are similar derivations for waves propagating over planar and curved two-dimensional surfaces respectively.

Let us start by considering wave fronts which are located by surfaces in $\mathbf{R}^3$ which propagate according to the rule given in (4.3.1).

If, at some instant of time, t, we are able to introduce local orthogonal curvilinear coordinates, η and λ, for points $\mathbf{r}$ on the surface, then N and K are given by

$$\begin{aligned} N &= \frac{\mathbf{r}_t . \mathbf{r}_\eta \times \mathbf{r}_\lambda}{|\mathbf{r}_\eta||\mathbf{r}_\lambda|}, \\ K &= -\frac{\mathbf{r}_{\eta\eta} . \mathbf{r}_\eta \times \mathbf{r}_\lambda}{|\mathbf{r}_\eta|^3|\mathbf{r}_\lambda|} - \frac{\mathbf{r}_{\lambda\lambda} . \mathbf{r}_\eta \times \mathbf{r}_\lambda}{|\mathbf{r}_\eta||\mathbf{r}_\lambda|^3}. \end{aligned} \tag{4.3.4}$$

Here the oriented normal defining the forward direction of the wave (where $u \approx 0$) is

$$\frac{\mathbf{r}_\eta \times \mathbf{r}_\lambda}{|\mathbf{r}_\eta||\mathbf{r}_\lambda|},$$

and $\times$ denotes the usual vector, or cross, product.

Suppose now we wish to restrict our attention to waves propagating over some oriented smooth surface S. For $\mathbf{r}^* \in S$, let $\mathbf{n}(\mathbf{r}^*)$ denote the oriented normal to S. Our wave front on S is located by a curve, say $\mathbf{r}^*(\eta, t)$, parameterised by η for each time t. Setting

$$\mathbf{r} = \mathbf{r}^*(\eta, t) + \lambda \mathbf{n}(\mathbf{r}^*(\eta, t))$$

in (4.3.4) is equivalent to seeking three-dimensional waves that intersect S orthogonally along $\mathbf{r}^*(\eta, t)$. Clearly, $\mathbf{r}_\lambda = \mathbf{n}$; so (4.3.1) and (4.3.4) simplify to yield an equation for the propagation of waves on the surface S, that is

$$\frac{\mathbf{r}_t.\mathbf{r}_\eta \times \mathbf{n}}{|\mathbf{r}_\eta|} = c + \frac{\varepsilon \mathbf{r}_{\eta\eta}.\mathbf{r}_\eta \times \mathbf{n}}{|\mathbf{r}_\eta|^3}$$

where $\mathbf{r} = \mathbf{r}|_{\lambda=0} = \mathbf{r}^* \in S$.

For example, if we wish to consider waves on the plane, we set

$$\mathbf{r} = \begin{bmatrix} x(\eta, t) \\ y(\eta, t) \\ \lambda \end{bmatrix}.$$

Clearly, for any time t, $\mathbf{r}_\eta \times \mathbf{r}_\lambda = 0$, and now (4.3.4) yields

$$\begin{aligned} N &= \frac{x_t y_\eta - y_t x_\eta}{(x_\eta^2 + y_\eta^2)^{1/2}} \\ K &= \frac{-x_{\eta\eta} y_\eta + y_{\eta\eta} x_\eta}{(x_\eta^2 + y_\eta^2)^{3/2}}. \end{aligned} \tag{4.3.5}$$

Thus we obtain the eikonal equation for waves in the plane [39].

If $\mathbf{r} = (\lambda + 1)\mathbf{e}(\eta, t)$ where $|\mathbf{e}| \equiv 1$, then we obtain

$$N = \frac{\mathbf{e}_t.\mathbf{e}_\eta \times \mathbf{e}}{|\mathbf{e}_\eta|}$$

and

$$K = \frac{-\mathbf{e}_{\eta\eta}.\mathbf{e}_\eta \times \mathbf{e}}{|\mathbf{e}_\eta|^3},$$

which defines the equation for front propagation on the unit sphere S_2.

If we assume for the moment that $\mathbf{r}_t.\mathbf{r}_\eta$ and $\mathbf{r}_t.\mathbf{r}_\lambda$ are zero, so that each point on the wave front (η, λ fixed) moves in the normal direction, we may rewrite (4.3.1) as

$$\mathbf{r}_t = [c - \varepsilon K] \frac{\mathbf{r}_\eta \times \mathbf{r}_\lambda}{|\mathbf{r}_\eta||\mathbf{r}_\lambda|}, \tag{4.3.6}$$

with K given by (4.3.4).

For waves on the plane, this becomes

$$x_t = [c - \varepsilon K].\frac{y_\eta}{(x_\eta^2 + y_\eta^2)^{1/2}},$$
$$y_t = [c - \varepsilon K].\frac{-x_\eta}{(x_\eta^2 + y_\eta^2)^{1/2}}, \tag{4.3.7}$$

while for waves on the sphere, we have

$$\mathbf{e}_t = \left(c + \varepsilon \frac{\mathbf{e}_{\eta\eta}.\mathbf{e}_\eta \times \mathbf{e}}{|\mathbf{e}_\eta|^3} \right) \frac{\mathbf{e}_\eta \times \mathbf{e}}{|\mathbf{e}_\eta|}. \tag{4.3.8}$$

(Notice: $\mathbf{e}.\mathbf{e}_t = 0$ so that $|\mathbf{e}(\eta, t)| \equiv 1$ for all η, t.) The equations (4.3.7) and (4.3.8) are parabolic systems which can, in theory, be solved to find the wave front position as an evolving function time.

Now let us concentrate on waves propagating in the (x, y)-plane. The equations contain a number of η-derivatives which tend to conceal the real simplicity of the problem. So we begin by considering some simplifications.

Firstly, suppose our curve is such that for each t fixed, $x(\eta, t)$ is invertible, with $x_\eta \leq 0$ say. Then writing $y^*(x, t) = y(\eta, t)$ where $x = x(\eta, t)$, we obtain

$$y_t = y_x^* x_t + y_t^*,$$

and

$$y_n = y_x^* x_\eta,$$

so that (4.3.7) implies

$$y_t^* = [c - \varepsilon K](1 + y_x^{*2})^{1/2}.$$

Now

$$y_{\eta\eta} = y_{xx}^* x_\eta^2 + y_x^* x_{\eta\eta},$$

so that

$$x_{\eta\eta} y_\eta - y_{\eta\eta} x_\eta = -x_\eta^3 y_{xx}^*.$$

Thus, substituting from (4.3.5) for K, and dropping the asterisks, we have

$$y_t = \varepsilon \frac{y_{xx}}{(1 + y_x^2)} + c(1 + y_x^2)^{1/2}. \tag{4.3.9}$$

Hence, if the *dummy* parameter η is replaced by the x-coordinate along the curve, (4.3.7) simplifies to (4.3.9), a remarkably simple quasi-linear evolution equation. Note first that $y = ct$ is a solution of (4.3.9), representing a plane wave propagating over the (x, y)-plane in the y-direction.

Linearizing (4.3.9) about $y = ct$, say,

$$y = ct + w(x,t)$$

it is easy to see that w satisfies

$$w_t = \varepsilon w_{xx}.$$

Thus, if $\int w(0,x)\,dx$ is finite, we have $w \to 0$ uniformly as $t \to \infty$: hence the plane wave solutions of (4.3.9) (and hence of (4.3.7)) are stable.

We shall refer to families of solutions of (4.3.6) which are stable (as a family) as being **geometrically stable**. The importance of this notion is that our geometric theory suggests that waves which take these forms may be stable solutions of reaction-diffusion systems. On the other hand, waves which are not geometrically stable are unlikely to be stable solutions of reaction-diffusion equations, particularly in the limit of a suitable constant ε being very small.

We have shown that the plane waves are geometrically stable. As a further example, change (4.3.7) to polar coordinates. If $x = r^*(\eta,t)\cos\theta(\eta,t)$, $y = r^*(\eta,t)\sin\theta(\eta,t)$ and $\theta(\eta,t)$ is invertible for fixed t, say $\theta_\eta > 0$: we may rewrite (4.3.7) as

$$r_t = \varepsilon\frac{[r_{\theta\theta} - \frac{2r_\theta^2}{r} - r]}{(r_\theta^2 + r^2)} + c\frac{(r_\theta^2 + r^2)^{1/2}}{r} \tag{4.3.10}$$

where $r(\theta,t) = r^*(\eta,t)$ whilst $\theta = \theta(\eta,t)$.

Simple solutions of (4.3.10) (and hence of (4.3.7)) are obtained by assuming r is independent of θ. Then,

$$r_t = c - \frac{\varepsilon}{r}. \tag{4.3.11}$$

Hence, we obtain a spherical wave, centred at the origin. Notice that if $r(0) > \varepsilon/c$, the wave spreads outwards with $r(t) \to \infty$ at $t \to \infty$. However, if $r(0) < \varepsilon/c$, then the wave collapses to $r = 0$ in finite time.

This is **threshold**-type behaviour. If not enough of the domain is initially stimulated, then the wave collapses.

Returning to the spherical waves, we have a three-parameter family of such waves, since we may specify $r(0)$ and, by a change of the coordinates, the centre of the wave.

In [24], it is shown that (4.3.1) admits spherical waves in $\mathbf{R}^3$ which are geometrically stable. We can obtain the same result here for spherical waves in $\mathbf{R}^2$ by linearizing (4.3.9) about a solution $R(t)$ of (4.3.11).

Let $r = R(t) + s(t,\theta)$ in (4.3.9), then linearizing, we have

$$s_t = \frac{\varepsilon}{R^2}[s_{\theta\theta} + s], \tag{4.3.12}$$

where $s(\theta) = s(2\pi + \theta)$.

The trivial solution $s = 0$ is unstable in the first eigenmode spanned by the function $\phi \equiv 1$. However, the perturbation in this mode can be assumed zero by resetting $R(0)$ if necessary.

The neutral stability in the second eigenmode spanned by the functions $\cos\theta$ and $\sin\theta$ corresponds to a shift in the origin of the spherical wave. Hence, without loss of generality, we may assume that at $t = 0$

$$\int_0^{2\pi} s\, d\theta = \int_0^{2\pi} s.\cos\theta\, d\theta = \int_0^{2\pi} s.\sin\theta\, d\theta \tag{4.3.13}$$

since otherwise, we can translate the origin or reset $R(0)$. The trivial solution of (4.3.11) is stable to perturbations satisfying (4.3.12). Hence, spherical waves in the plane are geometrically stable.

As we indicated earlier, we may utilize the equation (4.3.6) in situations where propagating waves meet the boundary of the domain $\Omega \subseteq S$ in which the reaction-diffusion system holds. (Recall that S is some smooth surface embedded in $\mathbf{R}^3$.)

In what follows, we shall always assume that appropriate no-flux (Neumann) boundary conditions have been imposed upon state variables of the reaction-diffusion system at boundaries $\partial\Omega$.

It is easy to see that our asymptotic solution of the reaction-diffusion system is valid only if waves which intersect the boundary do so orthogonally (i.e. at the wave-boundary intersection, $\mathbf{r}_\eta$ is always normal to the boundary). Then, by the symmetry of the asymptotic solution along the front, in the η-direction, we are assured that it satisfies the Neumann boundary condition.

Suppose at time $t = 0$, say, we parameterize our wave so that $\mathbf{r}(0,0) \in \partial\Omega$ (some smooth boundary curve), while $\mathbf{r}(\eta,0) \in \Omega$ for $\eta > 0$ in some interval. Then, for $t \geq 0$, $\mathbf{r}_\eta(0,t)$ must be normal to $\partial\Omega$, in the tangent plane to S (see Figure 4.6), so that $r_t(0,t)$ is tangent to $\partial\Omega$, and $\mathbf{r}(0,t)$ propagates along $\partial\Omega$.

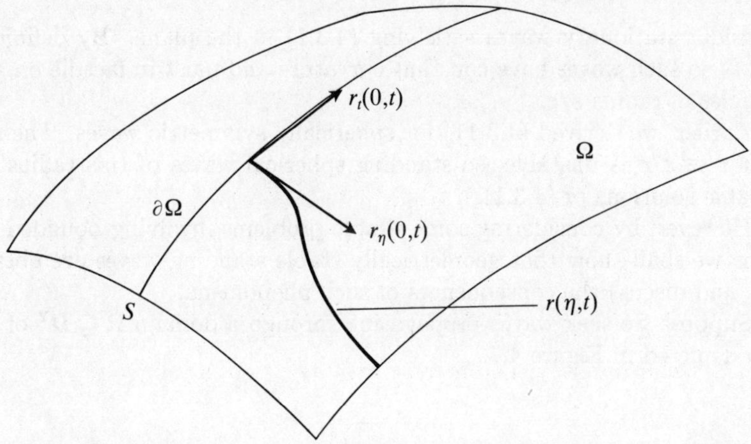

Figure 4.6: The wave meets $\partial\Omega$ in S.

We may state the boundary condition for the general situation, where $\Omega \subseteq S$, a smooth surface in $\mathbf{R}^3$, as follows:

Let $\partial\Omega = \{\mathbf{r} \in S : g(\mathbf{r}) = 0\}$ for some scalar field g, with $\nabla g \neq 0$, tangent to S on $\partial\Omega$. We require

$$\begin{aligned} g(\mathbf{r}(0,t)) &= 0 \\ \nabla g(\mathbf{r}(0,t)) \times \mathbf{r}_\eta(0,t) &= 0 \end{aligned} \qquad \text{for } t > 0. \tag{4.3.14}$$

Notice, our vector system (4.3.6) for waves propagating over surfaces is a pair of coupled second-order partial differential equations. Thus, if η varies in some finite interval, we must impose two conditions at each end point. Clearly, conditions of the form of (4.3.14) applied at each end point are commensurate with our equation. Hence, if we specify Ω with smooth boundaries, we may in principle solve (4.3.6) together with appropriate boundary conditions of the type (4.3.14) for waves propagating over Ω which meet $\partial\Omega$.

4.4 Stable stationary waves

Consider stationary waves satisfying (4.3.1) in the plane. By definition $N \equiv 0$, so such waves have constant curvature and must in fact lie on arcs of circles of radius ε/c.

Earlier, we derived (4.3.11) for spherically symmetric waves. The rest point $r \equiv \varepsilon/c$ is unstable, so standing spherical waves of this radius are unstable solutions of (4.3.1).

However, by considering some simple problems involving bounded domains we shall show that geometrically stable standing waves are obtainable, and discuss the consequences of such phenomena.

Suppose we seek waves propagating through a domain $\Omega \subseteq \mathbf{R}^2$ of the form depicted in Figure 4.7.

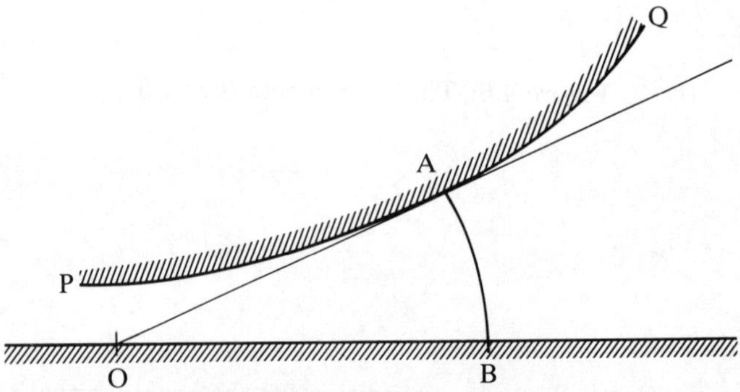

Figure 4.7: A stationary wave.

It is easy to construct such examples where a curve AB is an arc of a circle centred at the point O of radius ε/c, meets $\partial\Omega$ orthogonally. Clearly, the curve AB represents an equilibrium for (4.3.7) commensurate with the orthogonal boundary conditions introduced in section 4.3.

Let us introduce polar coordinates (r, θ) centred at the origin O. We assume the boundary curve PAQ is given by

$$g(r, \theta) = 0 \tag{4.4.1}$$

for some smooth function g, with nonvanishing gradient in a neighbourhood of A. Furthermore, let α^* denote the angle BOA, so that A has the coordinates $(\varepsilon/c, \alpha^*)$.

We choose to write (4.3.6) in polar coordinates with $r = r(\theta, t)$ as in section 4.3. Thus, perturbations of our stationary wave must satisfy

$$r_t = \varepsilon \frac{[r_{\theta\theta} - \frac{2r_\theta^2}{r} - r]}{(r^2 + r_\theta^2)} + c\frac{(r_\theta^2 + r^2)^{1/2}}{r} \tag{4.4.2}$$

for

$$\theta \in (0, \alpha), \quad t \geq 0,$$

together with boundary conditions

$$r_\theta(0, t) = 0, \tag{4.4.3}$$

$$g(r(\alpha, t), \alpha) = 0, \tag{4.4.4}$$

$$r^2(\alpha, t) g_r(r(\alpha, t), \alpha) = r_\theta(\alpha, t) g_\theta(r(\alpha, t), \alpha). \tag{4.4.5}$$

Here, $\alpha(t)$ depends on the solution r; so (4.4.2-5) is a moving boundary problem. The condition (4.4.3) ensures that the solution curve meets the x-axis orthogonally, while (4.4.4) and (4.4.5) ensure that the solution meets $g = 0$ orthogonally at some point $(r(\alpha, t), \alpha)$.

Now, $r = \varepsilon/c$ (and $\alpha = \alpha^*$) is by hypothesis a solution of (4.4.2-5). We perturb this by setting $r = \varepsilon/c + s$ and $\alpha = \alpha^* + \beta$, where both s and β are small. We linearize (4.4.2-5) to obtain

$$s_t = \frac{[s_{\theta\theta} + s].c^2}{\varepsilon}, \quad \theta \in [0, \alpha^*], \tag{4.4.6}$$

$$s_\theta(0, t) = 0, \tag{4.4.7}$$

$$0 = g_r s + g_\theta \beta, \quad \text{at} \quad \theta = \alpha^*, \tag{4.4.8}$$

$$(\varepsilon^2/c^2).g_{rr} s + g_{r\theta}\beta = s_\theta g_\theta, \quad \text{at} \quad \theta = \alpha^*, \tag{4.4.9}$$

where $g_r, g_\theta, g_{rr}, g_{r\theta}$ are evaluated at $(r, \theta) = (\varepsilon/c, \alpha^*)$.

However, since $r = \varepsilon/c$, and $\alpha = \alpha^*$ is an equilibrium, (4.4.5) implies $g_r(\varepsilon/c, \alpha^*) = 0$; hence, by (4.4.8), we have $\beta = 0$ and (4.4.9) becomes

$$(\varepsilon^2/c^2).(g_{rr}/g_\theta)s = s_\theta, \quad \text{at} \quad \theta = \alpha^*. \tag{4.4.10}$$

Here, we are assuming implicitly that $g_\theta \neq 0$ at $(r, \theta) = (\varepsilon/c, \alpha^*)$. If this were not the case, then we would have $\nabla g = 0$, which contradicts the definition of g. Eigenfunctions for the problem (4.4.8), (4.4.7), (4.4.10) are of the form

$$s = e^{(1-\lambda^2)t} \cos \lambda\theta, \tag{4.4.11}$$

where λ is real nonnegative and satisfies

$$-\lambda \tan \lambda\alpha^* = (\varepsilon^2/c^2).(g_{rr}/g_\theta)|_{(\varepsilon/c,\alpha^*)}. \tag{4.4.12}$$

Thus, if $0 < \alpha^* < \pi/2$, and our stationary solution is stable, we must have $\lambda > 1$ in (4.4.11), which, by (4.4.12), is true if and only if

$$(\varepsilon^2/c^2)(g_{rr}/g_\theta)|_{(\varepsilon/c,\alpha^*)} \quad \text{is not in } [-\tan\alpha^*, 0]. \tag{4.4.13}$$

Let us use an alternative form for the upper part of $\partial\Omega$ in Figure 4.7: that is

$$y = h(x), \quad \text{with } h' > 0,$$

where the origin in (x, y) coordinates is not necessarily coincident with our origin, O, introduced in relation to the polar coordinates. Then, if the x-coordinate of A is x^*, we have

$$h'(x^*) = \tan\alpha^*,$$

and

$$h(x^*) = (\varepsilon/c).\sin\alpha^*.$$

Thus

$$h(x^*) = (\varepsilon/c).\frac{h'(x^*)}{\sqrt{1+h'^2(x^*)}}. \tag{4.4.14}$$

Hence, our steady-state solution exists only if h satisfies (4.4.14) at some x^*.

The stability condition (4.4.13) may be written in terms of h using

$$g(r,\theta) = r\sin\theta - h(x^* - (\varepsilon/c)\cos\alpha^* + r\cos\theta).$$

Now using (4.4.13), we can easily show that either

$$(\varepsilon/c)h'' > h'(1+h'^2)^{3/2} \tag{4.4.15.a}$$

or

$$h'' < 0 \tag{4.4.15.b}$$

(at $x = x^*$) imply that the steady-state solution is locally stable.

Clearly, for concave domains such as in Figure 4.7, $h'' > 0$, so (4.4.15) say that if h'' is large enough at x^*, the wave is stable.

For example, suppose $h = \delta + Dx^2$ for $x > 0$, where δ is very small and D large. It can be shown that when δ is small, there are two possible standing waves, one where $x^* \approx \delta/2D$, and one for much larger x^*. The inner wave is locally stable while the outer wave is unstable.

As δ is increased, the stationary waves coalesce and vanish. Thus, if a wave is initiated at $x = 0,\ 0 \leq y \leq \delta$, for small δ, it tends towards the stationary wave intersecting the boundary at $x \approx \delta/2D$, within the domain $\Omega = \{x \geq 0 \text{ and } 0 \leq y \leq \delta + Dx^2\}$.

Having demonstrated one possible situation where a standing wave is stable, we pass over the opportunity to construct more elaborate examples and briefly focus upon the consequences of such behaviour.

The point to note is that reaction-diffusion waves cannot propagate out of thin slits. This principle can be used to provide a simple explanation of the result that homogeneous reaction-diffusion systems with two or more stable rest-states possess stable inhomogeneous steady state solutions in certain nonconvex domains. The classic example is a steady-state solution in a dumb-bell-shaped region, a narrow bridge between two large convex sub-domains. Here, the bridge is a small slit in our present terminology; thus, even though plane waves for the system propagate with nonzero speed, changing the state from one rest-state to another, it is easy to imagine that if the slit is narrow enough, it can be capped at one end by a circular arc which is a geometrically stable stationary wave-front configuration.

4.5 Spiral waves

Given a system of excitable reaction-diffusion equations (e.g. the FitzHugh-Nagumo equations), we wish to obtain spiral wave solutions similar to those exhibited in the Belousov-Zhabotinsky reaction. More precisely, we shall seek solutions which are of the form of a single spiral wave in the plane, rotating about the origin.

We would like to apply our eikonal equation directly to this problem; and since we know that such a wave ought to rotate uniformly about its central core, it is merely a matter of determining the geometry of the spiral. In order to use this approach, we must have a single *fast* variable, u, which dominates the wave-like behaviour (or excitability) of the system. For such a system, the analysis in section 4.3 is valid; ε is a parameter present explicitly in the equations, but the constant, c, is not so obviously determined. The point is that our system may possess many plane periodic waves (containing sharp transition-layers), each moving with different speeds. In Chapter 3, for example, we constructed such asymptotic solutions for the FitzHugh-Nagumo equations and sketched the dispersion curve which relates the speed of propagation to the wavelength. As $|\mathbf{r}| \to \infty$, our prospective spiral waves look like plane periodic waves (locally, at least), so some means must be found of choosing the correct wave speed, c, to be inserted into the eikonal equation, (4.3.1).

We begin by sketching out the general approach and resolving this last problem in the process. Once this is done, we make the calculations involved more explicit, and derive the equation which determines the geometry of our spiral wave.

We point out here that, strictly speaking, our solution will not be valid in some neighbourhood of the centre. This core region remains a problem. In the present analysis, we shall allow our transition-layer to reach as far as the origin, about which it rotates uniformly. However, the transition layer requires at least a width of $O(\varepsilon)$ in which to fill the corresponding transition profile. Hence, we expect that our solution will be valid to within an ε-sphere of the origin. Nevertheless, we are able to obtain some useful information regarding simple spiral waves in the plane. In particular, we shall show how their structure and angular velocity may be obtained uniquely. This is important as far as the Belousov-Zhabotinsky reaction is concerned because, although there may be many spirals in a single Petri dish, they all have the same (local) geometry, and all rotate with the same speed.

The approach presented in this section is essentially that of Keener [39], and Keener and Tyson [40]. Their analysis of spiral waves in the Belousov-Zhabotinsky reaction was one of the main motivating factors in developing the present geometrical theory for reaction-diffusion waves. In particular, it was their work that provided the impetus for the subsequent investigations in [24],[25].

Consider an anti-clockwise spiral wave, rotating clockwise about the origin (see Figure 4.8). We suppose that the wave possesses a transition layer located by such a spiral.

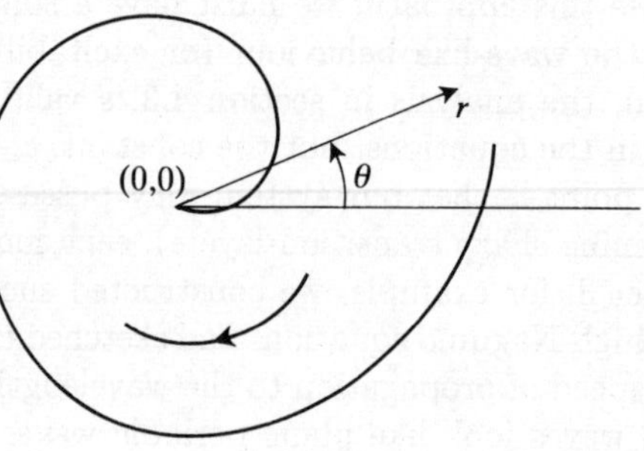

Figure 4.8: A spiral curve rotating about the origin.

Let us parameterize the curve locating the transition layer by $\eta \equiv r = \sqrt{x^2+y^2}$, so that in polar coordinates, the geometry and motion of the spiral is given by

$$\theta = g(r) - \omega t. \tag{4.5.1}$$

Here, the function g is to be determined, and must satisfy

$$g_r \geq 0, \tag{4.5.2}$$

$$\lim_{r\to\infty} g_r = g_\infty, \text{ constant}, \tag{4.5.3}$$

and is bounded at $r = 0$. Without loss of generality, we may set $g(0) = 0$. The condition on g, as $r \to \infty$, means that for large r, the waves are almost outgoing spherical waves, with wavelength

$$\lambda = \frac{2\pi}{g_\infty}.$$

The whole spiral satisfying (4.5.1-3) would have period $\frac{2\pi}{\omega}$, so, for large r, the transition layer would move a distance λ, in a radial direction, in time $\frac{2\pi}{\omega}$. Thus, an observer moving off to infinity would see waves moving with speed

$$\frac{\omega}{g_\infty}$$

asymptotically in a radial direction. For large enough r, the curvature of such waves becomes negligible, and our local observer would see plane periodic waves with speed ω/g_∞ and wavelength λ.

Now returning to our geometric theory where

$$c = N + \varepsilon K, \tag{4.5.4}$$

we must have

$$c = \frac{\omega}{g_\infty}$$

since ω/g_∞ is the speed of the *plane waves at infinity*. Hence,

$$g_\infty = c/\omega, \tag{4.5.5}$$

and

$$\lambda = 2\pi\omega/c. \tag{4.5.6}$$

Now, for c fixed, we must solve our eikonal equation (4.5.4) for the pair, (g, ω), satisfying (4.5.2-3) and (4.5.5). This is, in effect, a nonlinear eigenproblem.

Finally, we demand that the wavelength, λ, given by (4.5.6), lies on the dispersion curve, which is assumed known. Recall that the dispersion curve defines the relationship between the speed and the wavelength for plane periodic travelling waves. As c varies, we may regard ω as a functional of c, and hence require the curve (4.5.6) in the (λ, c)-plane to intersect the dispersion curve. The result is a unique intersection (or, at least, a finite number of intersections), determining c and hence ω.

In the examples we have in mind (such as the FitzHugh-Nagumo equations, see Chapter 3, Figure 3.12), the plane periodic waves possess sharp transition layers, so that, out at infinity, we imagine these to be wrapped up on to the spiral wave. (It may well be the case that such transition layers have already been of use in allowing us to get some idea as to how the dispersion curve behaves.)

In order to examine the geometry of the spiral waves, we use (4.5.1) and substitute

$$x(r) = r\cos(g(r) - \omega t)$$
$$y(r) = r\sin(g(r) - \omega t)$$

into the expressions (4.3.5), for N and K, which were derived for waves propagating in the plane. (Here we must identify the parameter η with r.) We have

$$\begin{aligned} N &= \frac{x_t y_r - y_t x_r}{(x_r^2 + y_r^2)^{1/2}} \\ &= \frac{r\omega}{(1 + r^2 g_r^2)^{1/2}} \end{aligned} \tag{4.5.7}$$

and

$$\begin{aligned} K &= \frac{-x_{rr} y_r + y_{rr} x_r}{(x_r^2 + y_r^2)^{3/2}} \\ &= \frac{2g_r + r g_{rr} + r^2 g_r^3}{(1 + r^2 g_r^2)^{3/2}}. \end{aligned} \tag{4.5.8}$$

Now, given c, we must solve $N + \varepsilon K = c$ to obtain the pair (ω, g), satisfying

$$g_r \geq 0, \quad \lim_{r\to\infty} g_r = \frac{c}{\omega}, \quad g(0) = 0.$$

If, for the moment, we assume that the εK term is negligible, we obtain

$$g_r = \left(\frac{\omega^2}{c^2} - \frac{1}{r^2}\right)^{1/2} \tag{4.5.9}$$

whose solution is defined for $r \geq c/\omega$. In fact, (4.5.9) may be solved explicitly as follows:

$$g(r) = \left(\frac{r^2\omega^2}{c^2} - 1\right)^{1/2} - \tan^{-1}\left(\frac{r^2\omega^2}{c^2} - 1\right)^{1/2}.$$

Clearly, $g_r \to c/\omega$ as $r \to \infty$, but g is unable to satisfy the boundary condition at $r = 0$, since it cannot reach this far. The curve represented by the solution of (4.5.9) is the involute of a circle of radius c/ω. (This is the curve traced by the end of a taut string unwinding from a cylindrical post.)

The similarity between such involute spirals and the spiral waves observed in excitable media has been of interest for some time. However, their deficiencies have been well documented [77], in particular, their failure to reflect the effects of curvature on normal velocity, [79], and lack of agreement with experiments near to the centre, or core, of the spiral.

In our present analysis, we must reject the involute spiral on three major counts:
(i) It fails to provide any explanation of the core geometry, and is undefined for small r.
(ii) It fails to fix the parameter ω in terms of c. (Recall that this was important in order to force the *plane waves at infinity* to satisfy the dispersion curve, and hence possibly to obtain unique spirals.)
(ii) As $r \to c/\omega$, $g_{rr} \to \infty$. Hence, when K, in (4.5.8), is evaluated for the involute, it becomes unbounded, as $r \to c/\omega$. Thus, the εK term is non-negligible here, and we should solve the full eikonal equation rather than $N = c$.

To proceed, we must solve the eikonal equation. However, since we require $g_r \to c/\omega$ as $r \to \infty$, we will have $K \to 0$ (corresponding to the locally plane waves at infinity). Thus, as $r \to \infty$, we obtain $c = N$, so our solution must be asymptotic to an involute spiral for large r.

We have

$$c = \varepsilon \frac{(2g_r + rg_{rr} + r^2 g_r^3)}{(1 + r^2 g_r^2)^{3/2}} + \frac{\omega r}{(1 + r^2 g_r^2)^{-1/2}}. \tag{4.5.10}$$

As $r \to 0$, we require $g \to 0$, so we must have $g_r \to c/2\varepsilon$.

Setting $h = rg_r$ in (4.5.10), we obtain

$$rh_r = (1 + h^2)\left(\frac{rc}{\varepsilon}(1 + h^2)^{1/2} - \frac{\omega r^2}{\varepsilon} - h\right) \tag{4.5.11}$$

together with the boundary conditions

$$\begin{aligned} h(0) &= 0, \\ \lim_{r \to \infty} h' &= \frac{c}{\omega}. \end{aligned} \tag{4.5.12}$$

At $r = 0$, we must have $h_r = c/2\varepsilon$, so that (4.5.12) is compatible with (4.5.11). We can numerically solve (4.5.11) together with

$$h(0) = 0, \quad h_r(0) = \frac{c}{2\varepsilon} \tag{4.5.13}$$

for different values of ω. As ω varies, the solution in the phase plane displays different behaviour as $r \to \infty$. For precisely one value of ω, h_r remains bounded, and tends to c/ω. The corresponding orbit may be integrated to provide our solution g. For our current purposes, we are interested in ω, which has been determined by the above calculation.

For different choices of c, we can determine the corresponding values for ω, and write $\omega = \omega(c)$.

Now, we sketch the curve $\lambda = 2\pi\omega/c$ given in (4.5.6) in the (c, λ)-plane. Superimposing the dispersion relation, assumed known (see Figure 3.12, for example), we obtain an intersection corresponding to a uniquely determined value of c, and hence of ω, for spiral waves.

This programme was carried out for the Belousov-Zhabotinsky reaction, [40]. We refer those interested to [39] and [40].

Having seen how the geometry of the spiral wave itself is able to fix the speed of rotation, we move on to consider some other wave-like structures.

In [25], an analysis of propagating waves on spherical surfaces in $\mathbf{R}^3$ is presented. In particular, it is possible to analyse waves which connect *core* regions at the north and south poles, and rotate uniformly about a north-south pole axis. Such waves can be thought of as double spirals, one at each pole; see Figure 4.9.

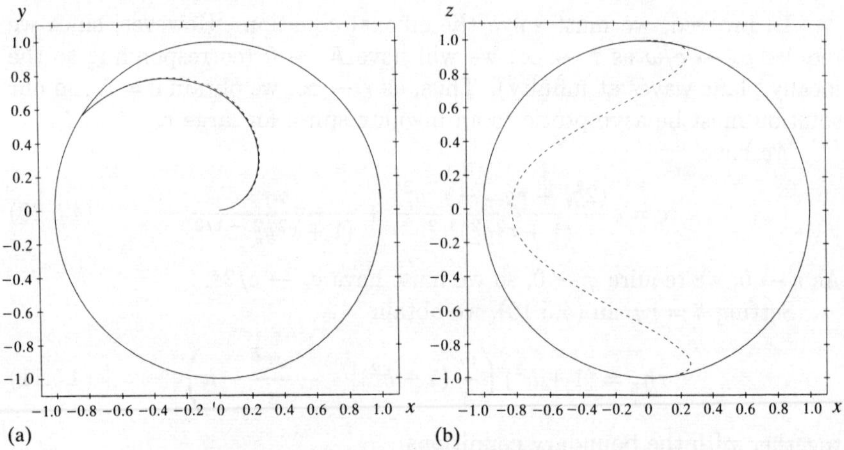

Figure 4.9: Rotating waves on the unit sphere.

In section 4.3, we have already derived the expressions for N and K evaluated for curves, $\mathbf{e}(\eta, t)$, on the unit sphere. Figure 4.9 really represents a solution of the resulting eikonal equation after substituting

$$\mathbf{e}_t = \mathbf{e} \times \mathbf{k}\omega$$

which imposes a uniform rotation about the north-south pole axis. It is worth remarking that recent laboratory experiments (at the time of writing) have shown such waves to exist, almost two years after their theoretical prediction [25].

Box J: The Belousov-Zhabotinsky reaction

In his book [77] Winfree provides a compelling account of the behaviour exhibited by excitable systems such as the Belousov-Zhabotinsky reaction. He has also recorded the following story behind its discovery.

Around 1950, Boris Belousov was a biochemist at Moscow State University. He was interested in cyclic phenomena, such as the Krebs cycle (which describes how organic carbon chains may be oxidized to produce carbon dioxide which return to the original compound via the formation of citric acid). Belousov created an inorganic analogue. He used cerium to oxidize citric acid in the presence of bromate. The solution cycled in time with a period of a few minutes. The cycles could be observed by the appearance and disappearance of the yellow oxidized cerium ions.

Between 1951 and 1957, Belousov tried to make his experiments and observations more precise, but he was unable to publish any detailed papers. The problem was that the prevailing opinion amongst chemists at the time was that solutions of chemical reagents could not oscillate; instead, the course of such reactions simply tended towards a final steady state.

Belousov lost the argument, and lost heart. However, his reagent became known amongst Moscow chemists, arousing cautious interest. In 1961, a graduate student called Zhabotinsky began to make his own investigations with Belousov's reagent. He replaced the citric acid with malonic acid, and made a number of refinements.

Throughout the 1960's, interest in cyclic reactions grew (and others were discovered). In 1970, Belousov died. His work had been vindicated by the more modern research. He was awarded the Lenin prize in 1980.

The cyclic behaviour of the Belousov-Zhabotinsky reagent is not the end of the story. In each cycle, the reagent lingers in a state quite close to steady state. After a while, the reaction gathers pace and the cycle begins again. By tinkering with the recipe, [77], these quiescent stages can be made to last longer and longer, up until the steady state becomes locally stable. When this happens,the reaction is excitable rather than cyclic. If the quiescent steady-state is perturbed sufficiently, the reaction moves through an excursion, similar to the previously described cycles, before returning to the stable state.

A qualitative example is provided by the system:

$$\begin{aligned} u_t &= f(u) - v, \\ v_t &= \varepsilon g(u, v), \end{aligned} \tag{1}$$

where ε is sufficiently small, and f and g are depicted in Figure 4.10. As the intersection of the nullclines moves from the middle branch of $v = f(u)$, to either outer branch, the system changes from being oscillatory (with a stable limit cycle), to excitable (with a locally stable rest point).

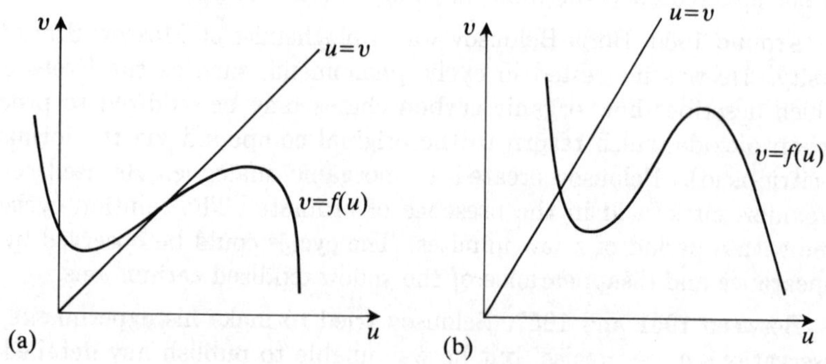

Figure 4.10: Nullclines for (1): (a) Oscillatory; (b) Excitable.

4.6 Toroidal scroll waves

Here, we shall use the eikonal equation to discuss some properties of toroidal scroll waves. Experimental evidence suggests that excitable systems of reaction-diffusion equations may support such solutions in three-dimensional domains (see Figure 4.2). We shall look for surfaces with rotational symmetry satisfying (4.3.1), with a singular filament in the form of a circle, following [24].

Firstly, we must introduce a suitable coordinate system. This is generated by rotating a polar coordinate system located at a distance R_0 from the z-axis, about the z-axis. We have

$$\begin{aligned} x =& (R_0 + \rho\cos\psi)\cos\phi \\ y =& (R_0 + \rho\cos\psi)\sin\phi \\ z =& -\rho\sin\psi \end{aligned} \tag{4.6.1}$$

where (x, y, z) are the usual Cartesian coordinates, so that $\mathbf{r} = \mathbf{r}(\rho, \phi, \psi)$ (see Figure 4.11).

Since we are interested in surfaces with rotational symmetry about the z-axis, we set

$$\lambda = \rho,\ \eta = \phi,\ \psi = \psi(\rho, t),$$

(recall η and λ were introduced earlier and are general parameters for our surface).

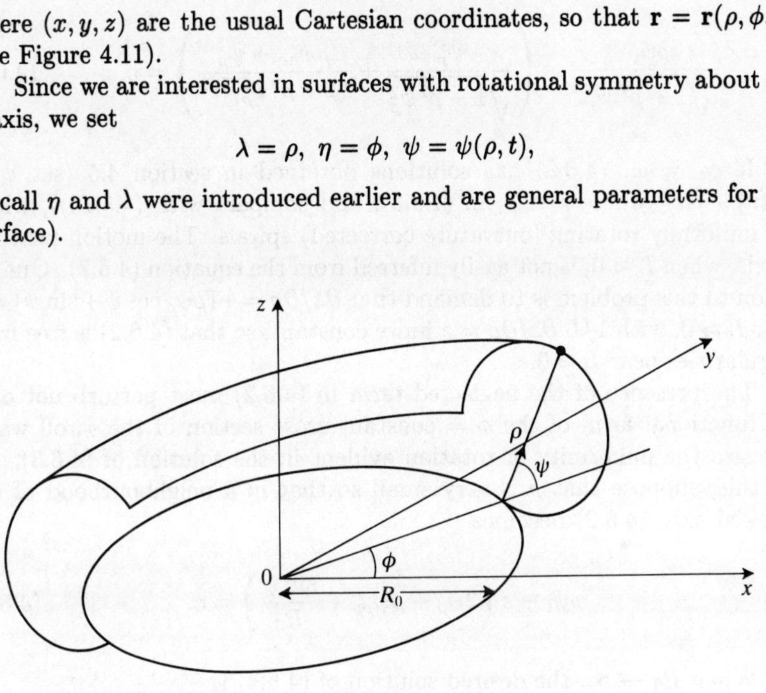

Figire 4.11: Toroidal coordinates and a toroidal scroll.

Now using (4.6.1), we substitute the corresponding derivatives of $\mathbf{r}$ into (4.3.4), so that $N + \varepsilon K = c$ becomes

$$\begin{aligned} \frac{\rho\psi_t}{\sqrt{1+\rho^2\psi_\rho^2}} - \left(\frac{\psi_\rho}{\sqrt{1+\rho^2\psi_\rho^2}} + \frac{\rho\psi_{\rho\rho}+\psi_\rho}{(1+\rho^2\psi_\rho^2)^{3/2}}\right)\varepsilon \\ -\frac{\varepsilon}{L}\frac{\sin\psi + \rho\psi_\rho\cos\psi}{\sqrt{1+\rho^2\psi_\rho^2}} = c, \end{aligned} \tag{4.6.2}$$

where $L \equiv R_0 + \rho \cos \psi$.

Notice that since our surface is defined by $\psi(\rho, t)$, it necessarily possesses rotational symmetry, and ρ and ϕ are (trivially) orthogonal coordinates.

Ideally, one should seek a solution satisfying $\psi(\rho, t) + 2\pi = \psi(\rho, t + 2\pi/\omega)$ for all t and all ρ, for some constant ω – that is, obtain an oscillatory structure with period $2\pi/\omega$. However, for $R_0 \gg \rho$ and $\varepsilon/R_0 \ll 1$, we can neglect the third term on the left-hand side and arrive at

$$\frac{\rho\psi_t}{\sqrt{1+\rho^2\psi_\rho^2}} - \varepsilon\left(\frac{\psi_\rho}{\sqrt{1+\rho^2\psi_\rho^2}} + \frac{\rho\psi_{\rho\rho}+\psi_\rho}{(1+\rho^2\psi_\rho^2)^{3/2}}\right) = c. \tag{4.6.3}$$

If $\psi_t = \omega$, (4.6.3) has solutions obtained in section 4.5 (set $\psi = -g(r) + \omega t$), where $r \equiv \rho$, in (4.6.3), and compare with (4.5.10)), which are uniformly rotating (curvature corrected) spirals. The motion near the z-axis, when $L \approx 0$, is not easily inferred from the equation (4.6.2). One solution to this problem is to demand that $\partial z/\partial\rho = -(\rho\psi_\rho \cos\psi + \sin\psi) = 0$ near $L \approx 0$, with $1/L\ \partial z/\partial\rho =$ a finite constant, so that (4.6.2) is free from singularities near $L \approx 0$.

The presence of the neglected term in (4.6.2) must perturb not only the functional form of the $\phi =$ constant cross section of the scroll wave, but also the uniformity of rotation evident in the solution of (4.6.3). To see this, suppose that ρ is very small so that in a neighbourhood of the toroidal axis, (4.6.2) becomes

$$\rho\psi_t - \varepsilon\left(2\psi_\rho + \rho\psi_{\rho\rho} + \frac{\sin\psi}{R_0}\right) = c. \tag{4.6.4}$$

When $R_0 \to \infty$, the desired solution of (4.6.4) is

$$\psi = \psi_0 = \omega t - \frac{c\rho}{2\varepsilon} + O(\rho^2),$$

which represents an expansion, for small ρ, of our spiral solution of (4.6.3).

Now substituting

$$\psi = \psi_0 + \frac{1}{R_0}\psi_1 + O\left(\frac{1}{R_0^2}\right)$$

into (4.6.4), we obtain

$$\psi_1 = -\frac{\varepsilon}{c}\cos\omega t + O(\rho).$$

Thus

$$\psi = \omega t - \frac{c\rho}{2\varepsilon} + O(\rho^2) + \frac{1}{R_0}\left(-\frac{\varepsilon}{c}\cos\omega t + O(\rho)\right) + O\left(\frac{1}{R_0^2}\right)$$

defines a surface for ρ small, which rotates in a nonuniform manner. (Nevertheless, it is still $2\pi/\omega$-periodic, since $\psi(\rho, t + 2\pi/\omega) = \psi(\rho, t) + 2\pi$).

A full analysis of toroidal scrolls is still awaited. It is not a very straightforward matter to find time-periodic solutions for (4.6.2), particularly since we must impose the smoothness condition

$$\frac{\partial z}{\partial \rho} = 0$$

whenever the wave meets the z-axis.

We may make similar constructions to those above in considering uniformly twisted toroidal scrolls. Here ψ must depend linearly on ϕ (with integer coefficient), so the corresponding equations are much more complex.

5 Nonlinear dispersal mechanisms

5.1 Introduction

This chapter is concerned with a number of models which utilize nonlinear advection and nonlinear diffusion terms so as to generate solutions exhibiting patterned structure. In Chapter 2, we saw how the coupling of (nonlinear) reaction terms and (linear) diffusion terms led to the development of structure in the long-time behaviour of solutions. The mechano-chemical models introduced in section 2.4 took this a stage further by coupling the variables through nonlinear flux terms. Thus the patterns in Figures 2.14 and 2.15 are internally *flux-driven*, and this is the kind of behaviour that we shall concentrate on in this chapter.

Of course, the mechano-chemical models represent fairly complicated systems (they were introduced in Chapter 2 to illustrate patterns with both spatial and temporal oscillations), and one may observe patterns driven by nonlinear transport mechanisms in much simpler circumstances.

We begin with an example.

In section 1.2, we derived the Fokker-Planck equation for a population of individuals, each making a random walk biased by some velocity field. A simple feedback transport mechanism is obtained by making the velocity field dependent upon the density function itself.

Consider a population of insects say, with density distribution $u(x,t)$, which *swarm* according to the following simple rule:

$$u_t = \varepsilon u_{xx} - (uw)_x, \quad x \in \mathbf{R}, \quad t \geq 0 \tag{5.1.1}$$

$$w = \int_x^\infty u(y)\,dy - \int_{-\infty}^x u(y)\,dy. \tag{5.1.2}$$

(5.1.1) is just the one-dimensional Fokker-Planck equation (see Box A). The velocity field w given by (5.1.2) advects individuals towards the *centre* of the distribution since, at each x, w is positive (respectively negative) if there are more individuals located to the right (respectively left) of x.

Hence we expect that solutions of (5.1.1-2) may develop a peak in the centre as individuals aggregate.

Notice first that the total mass (assumed finite),

$$M = \int_{-\infty}^\infty u(y)\,dy,$$

is constant for the motion given by (5.1.1-2).

Introduce the primative

$$v = \int_{-\infty}^{x} u(y)\,dy$$

so that (5.1.1-2) may be rewritten as

$$v_t = \varepsilon v_{xx} - v_x(M - 2v), \tag{5.1.3}$$

where

$$v \to 0 \quad \text{as} \quad x \to -\infty$$
$$v \to M \quad \text{as} \quad x \to +\infty.$$

The steady-state solution, v_0 say, for (5.1.3) satisfies

$$\varepsilon v_x = \frac{1}{2} v(M - v)$$

and is determined up to a rigid translation in x. In fact

$$v_0(x) = \frac{M}{2}\left[1 + \tanh\left(\frac{M}{4\varepsilon}(x - x_0)\right)\right] \tag{5.1.4}$$

for any constant x_0. Thus, the corresponding steady state, u_0, for (5.1.1-2) is given by

$$u_0(x) = \frac{M^2}{\varepsilon 8} \operatorname{sech}^2\left(\frac{M}{4\varepsilon}(x - x_0)\right).$$

Note that, as $\varepsilon \to 0$, u_0 becomes more sharply peaked about $x = x_0$, and approaches a point mass in the limit. This is illustrative of the fact that the diffusion term is necessary in order that solutions of (5.1.1-2) stay within a smooth class of functions. The existence theory of section 1.6 may be applied to (5.1.1-2) to show that solutions to the initial-value problem exist in $L_p(\mathbf{R})$ ($p = 1, 2$) provided $\varepsilon > 0$. If $\varepsilon = 0$, and we consider (5.1.3) subject to $v(x, 0) = \tilde{v}(x)$, we obtain a solution (by the method of characteristics, see either [22] or Chapter 3) of the form

$$v = \tilde{v}(s)$$

where

$$0 = s - x + t(M - 2\tilde{v}(s)).$$

Hence v develops a jump discontinuity (shock) at time

$$t = \inf_{s \in \mathbf{R}} \left\{ \frac{1}{2\frac{d\tilde{v}}{ds}(s)} \right\}.$$

Clearly, as $v(x,t)$ loses continuity and develops a jump, $u(x,t)$ must develop a point mass.

The interest in nonlinear transport mechanisms has been fairly varied. We shall discuss a few applications here before examining some models in more detail later in the chapter.

There has long been an interest in *long-range* dispersal phenomena in population and cell biology. The idea here is that the dispersal of particles is functionally dependent upon the density over a wide range compared to Fickian diffusion which determines the flux as a consequence of a single differentiation.

Let us consider a one dimensional transport process given by

$$\begin{aligned} u_t &= -\, J_x \\ J &= -\int k(y) u_x(x+y,t)\, dy. \end{aligned}$$

Here, the kernel k tends to zero away from the origin. Thus the flux J is a kind of local average of the usual Fickian flux $-u_x$.

Expanding $u_x(x+y,t)$ as a Taylor series, we obtain

$$\begin{aligned} J = -u_x(x,t) \int k(y)\, dy - u_{xx}(x,t) \int y k(y)\, dy \\ - u_{xxx}(x,t) \int \frac{y^2 k(y)}{2}\, dy - \dots. \end{aligned}$$

The odd moments $\int y^{2p+1} k(y)\, dy$ are often assumed to be zero on the grounds of symmetry but all the even moments may be nonzero. If one truncates the series for J, our model is approximated by a partial differential equation.

However, this truncated approach is inappropriate where the state variable represents a concentration or density function, and must therefore remain nonnegative. For example, suppose we set

$$-J = d_1 u_x + d_2 u_{xxx}$$

for some real constants d_1, d_2. Then our model becomes

$$u_t = d_2 u_{xxxx} + d_1 u_{xx}$$

for $x \in \mathbf{R}$, say, and $t > 0$.

Now, if $d_2 > 0$, perturbations with large wave numbers will grow arbitrarily quickly (take the Fourier transform), which suggests ill-posedness. On the other hand, if $d_2 < 0$, then the fourth-order term is smoothing (perturbations with large wave numbers decay quickly) and there is no mechanism for pattern generation, unless $d_1 < 0$. However, in this case, if u is nonnegative and agrees with $(x - x_0)^2$ in a neighbourhood of $x = x_0$ say, then $u_{xxxx} = 0$ at $x = x_0$, and so $u_t < 0$ there, which means u will become negative. This higher order term has been utilized previously, for example, to provide additional smoothness, although such *long range diffusion* seems rather delicate (the kernel k above must be negative for intermediate y values).

As suggested by our earlier example, an approach based on the Fokker-Planck equation is more fruitful. In section 5.2, we discuss chemotaxis which is one such application. There, the motion of individual cells is mediated by the secretion and decay of a chemical attractant. A number of similar applications are apparent.

Consider a solution containing charged chemical ions. Let u_i denote the concentration of the ith species. Then we have

$$u_{i\ t} = D_i \Delta u_i - \nabla . [u_i \alpha_i z_i \nabla \phi] + R_i$$

where D_i is the molecular diffusivity, R_i is the source/sink of ions due to chemical reaction, z_i denotes the charge of ions of the ith species, α_i is the mobility of ions of the ith species, and ϕ is the electrical potential field. ϕ must be derived from Poisson's equation:

$$\Delta \phi = \sum_{i=1}^{N} z_i u_i .$$

Thus the concentrations are coupled through the advection terms (via Poisson's equation) as well as through the reaction terms.

A similar situation arises in models for the distribution of electrons and holes in semi-conductors, as well as space-charge problems concerning the distribution of electrons in a vacuum.

Even in the absence of reaction terms, such models are complicated and the time-dependent behaviour may be difficult to discuss numerically. Fortunately, there are standard numerical packages for solving Poisson's equation on simple domains, with simple linear boundary conditions, so that small systems may be analysed numerically. The figures depicting solutions for our reduced mechano-chemical model (2.4.6-10) were obtained in that manner.

In section 5.3 we consider some models for clustering behaviour in population biology. The existence of solutions displaying a high degree of localized spatial structure is discussed. We also consider the travelling front problem for a scalar model of this type (representing the spread of a single species).

Thus far this book has concentrated upon linear (Fickian) diffusion processes. In section 5.4 we take this a stage further and consider some nonlinear diffusion mechanisms. A number of applications are given and we describe the qualitative properties of solutions. In fact such solutions are usually compactly supported and are **weak** ones (they do not possess the derivatives specified in the equations at the boundaries of their supports). We defer a discussion of weak solutions until section 5.5, where we also discuss some properties of travelling front solutions for reaction-(nonlinear) diffusion problems.

Before ending this section, we introduce the Navier-Stokes equations in Box K. These contain nonlinear advection terms, and are discussed partly for completeness, and partly because the *slow flow* approximation is used in section 5.4.

Although most of the analysis developed in this book was originally aimed at semilinear parabolic models in chemistry and biology, such methods have also yielded some new insights into more traditional areas of applied mathematics, such as fluid mechanics. Rather than trying to play down nonlinear effects, so as to gain mathematically amenable problems, we should attack them with every means; qualitative analysis, asymptotic analysis, the desktop P.C. and the largest supercomputer!

Box K: Incompressible fluid flow

Among the most well-known semi-linear systems are the Navier-Stokes equations. These are a classical model for the flow of an incompressible viscous fluid and take the form

$$\begin{aligned} u_t + uu_x + vu_y + wu_z &= -p_x + \frac{1}{Re}\Delta u + F_1, \\ v_t + uv_x + vv_y + wv_z &= -p_y + \frac{1}{Re}\Delta v + F_2, \\ w_t + uw_x + vw_y + ww_z &= -p_z + \frac{1}{Re}\Delta w + F_3, \\ u_x + v_y + w_z &= 0. \end{aligned} \tag{1}$$

Here, $\mathbf{u} = (u, v, w)^T$ is the velocity field with respect to the Cartesian coordinates (x, y, z), p denotes the pressure, $\mathbf{F} = (F_1, F_2, F_3)^T$ denotes the acceleration due to any externally applied body forces (gravity, for example), and *Re* is a parameter known as the Reynolds number. Assuming that the problem to be considered contains typical scalings of the form

$$\|\mathbf{u}\| = U,$$
$$\|\mathbf{x}\| = L,$$

we have $Re = \rho UL/\mu$, where ρ is the fluid density (constant) and μ is the fluid viscosity. In vector form, we have

$$\mathbf{u}_t + (\mathbf{u}.\nabla)\mathbf{u} = -\nabla p + \frac{1}{Re}\Delta\mathbf{u} + \mathbf{F} \qquad (2)$$
$$\nabla.\mathbf{u} = 0. \qquad (3)$$

The pressure, p, is like the tension in an inextensible string and assumes whatever behaviour is necessary in order to *hold the system together*, which in this case means allowing the fluid to remain incompressible. In Box B we perceived a pressure-like variable as a Lagrange multiplier associated with a constraint of the form (3).

The nonlinear term $(\mathbf{u}.\nabla)\mathbf{u}$, which appears in the material time derivative on the left-hand side, means that the existence and construction of solutions are nontrivial problems (see [35] for existence, using the methods developed in Chapter 1).

Due to wide applicability, various simplifying assumptions are often made in order to analyse problems explicitly. Principal among these is the **slow flow** model derived in the limit $Re \to 0$. Here, we rescale the pressure

$$\tilde{p} = \frac{p}{Re}$$

so that, in the absence of body forces, we can retain only those terms $O(\frac{1}{Re})$ in (2). We obtain

$$\nabla\tilde{p} = \Delta\mathbf{u}$$
$$0 = \nabla.\mathbf{u}.$$

This is the slow flow model,representing a pseudo-equilibrium approach to (2) (3) (in the sense that $d\mathbf{u}/dt \approx 0$).

Returning to (2), the terms $(-\nabla p + \frac{1}{Re}\Delta\mathbf{u})$ are derived by writing the internal body force in the form of the divergence of the **stress tensor**, $\underline{\sigma}$, where

$$\underline{\sigma} = -pI + \mu[\nabla\mathbf{u} + \nabla\mathbf{u}^T].$$

Thus we have

$$\nabla.\underline{\sigma} = -\nabla p + \mu\Delta\mathbf{u},$$

and (2) is sometimes written as

$$\rho(\mathbf{u}_t + (\mathbf{u}.\nabla)\mathbf{u}) = \nabla.\underline{\sigma} + \mathbf{F},$$

(abusing our notation to allow **F** to denote the external body forces).

This formulation is useful since the boundary conditions on fluid surfaces often take the form of specification of the normal and tangential stresses. The oil-drop problem in section 5.4 is an example of such a problem.

There are many excellent texts on fluid and continuum mechanics which discuss the Navier-Stokes equations and their applications [37],[74].

5.2 Chemotaxis

In the early stages of cartilage formation, patterned aggregations of mesenchymal cells are formed. These aggregations undergo various morphogenetic transformations as the limb takes shape. In the model here, we suppose the cells move towards a source of chemoattractant, which the cells themselves secrete. There are several well-known examples of cell aggregation arising from the cells secreting their own chemoattractant, such as in the aggregation phase of the slime mould *Dictyostelium discoideum*, where motile cells move towards a chemoattractant that is thought to be secreted by a small group of *pacemaker* cells (see, for example, [43], [69]). The analysis in this section follows [28].

The model system consists of two conservation equations for the concentrations of the cells, u, and the chemoattractant, c. We take the equations for u and c, proposed in [53], namely

$$\begin{aligned} u_t =& M\Delta u - a\nabla.(u\nabla c) \\ c_t =& D\Delta c + bu/(u+h) - \mu c \end{aligned} \tag{5.2.1}$$

where $M > 0$ and $a > 0$ are the cell motility and chemotactic parameters respectively, and D is the diffusion coefficient of c. The secretion rate is taken to be a typical so-called Michaelis-Menten saturating function, $bu/(u+h)$, with b and h positive, and a first-order kinetics degradation rate for c, namely μc, is assumed.

We nondimensionalize the system by introducing a typical time scale, $T = 1/\mu$, based on the chemoattractant decay time, a concentration $C = b/\mu$, that is based on the maximum secretion rate by the cells, and a length $L = (ab)^{1/2}/\mu$. We thus define the following dimensionless quantities:

$$\begin{aligned} u^* =& u/h, \quad c^* = c\mu/b, \quad x^* = x\mu/(ab)^{1/2}, \\ t^* =& \mu t, \quad D^* = D/ab, \quad M^* = M/ab, \end{aligned} \tag{5.2.2}$$

with which (5.2.1) becomes, after dropping the asterisks for notational convenience,

$$\begin{aligned}\partial u/\partial t &= M\Delta u - \nabla . u\nabla c,\\ \partial c/\partial t &= D\Delta c + u/(1+u) - c.\end{aligned} \tag{5.2.3}$$

We have thus reduced the parameters to the two dimensionless parameter groupings D and M in (5.2.3). In this section, we consider the nonlinear spatially heterogeneous steady-state solutions of (5.2.4) and investigate the possible bifurcations.

We begin by considering the time dependent problem for (5.2.4) in one space dimension, namely

$$\begin{aligned}u_t &= Mu_{xx} - (uc_x)_x\\ c_t &= Dc_{xx} - c + g(u)\end{aligned} \qquad x \in (0,1),\ t > 0 \tag{5.2.4}$$

where $g(u) = u/(1+u)$. Considering the biological situation, we shall impose no-flux or periodic boundary conditions. For definiteness, we choose here the former, that is

$$0 = u_x = c_x \ \text{ at } \ x = 0, 1 \ \text{ for } \ t > 0. \tag{5.2.5}$$

Our results will easily generalize to problems defined on compact domains in $\mathbf{R}^n$, with Neumann boundary conditions. However, our subsequent analysis in section 5.3 may not be directly extended, since we shall rely on a time-map argument which exploits the fact that x is a scalar variable. Also, if (5.2.5) were replaced by periodic boundary conditions, all the results here would remain valid with only minor adaptations. Thus we shall see that the existence and bifurcation of steady-state solutions for (5.2.4) with (5.2.5) naturally infers the existence of steady state solutions for (5.2.4) defined for $x \in \mathbf{R}$, with period 2.

Firstly, we consider uniform equilibrium solutions for (5.2.4), (5.2.5). If $u = u(x) \geq 0$, $c = c(x) \geq 0$ satisfy (5.2.4)-(5.2.5), then (u, c) is a steady-state or equilibrium solution and we define the norms

$$h(u) = \max_{x\in[0,1]} |u_x|$$

and

$$N(u) = \int_0^1 u(x)\, dx.$$

Here, $N(u)$ is the usual L_1 norm and, since u is a nondimensional cell density, it is the total number of cells present. For the time-dependent problem $dN(u)/dt = 0$.

The norm $h(u)$ is the usual ∞-norm of $u_x(x)$. Clearly, $h(u) = 0$ implies both u and c are spatially uniform.

Next, we shall show that for all constants $u_0 > 0$, there exists a uniform steady-state solution for (5.2.4),(5.2.5) of the form

$$(u, c) = (u_0, g(u_0)). \tag{5.2.6}$$

Such a solution is locally stable if and only if

$$M(Dn^2\pi^2 + 1) - u_0/(1 + u_0)^2 \tag{5.2.7}$$

is strictly positive for $n = 1, 2, \ldots$. Moreover, if (5.2.7) is stictly negative for some integer n, then the solution is unstable in the corresponding eigenmode ($\cos nx$). To see this, we linearize (5.2.4) about (5.2.6), and set

$$\begin{aligned} u =& u_0 + a(t) \cos n\pi x \\ c =& g(u_0) + b(t) \cos n\pi x \end{aligned}$$

for any $n = 1, 2, \ldots$. We obtain

$$\frac{d}{dt}\begin{bmatrix} a \\ b \end{bmatrix} = \begin{bmatrix} -\pi^2 n^2 M & u_0\pi^2 n^2 \\ (1+u_0)^{-2} & -D\pi^2 n^2 - 1 \end{bmatrix} \begin{bmatrix} a \\ b \end{bmatrix} \tag{5.2.8}$$

(since $\cos n\pi x$ is the nth eigenfunction of the Laplacian defined on (0,1) with Neumann boundary conditions).

The result follows immediately by considering the sign of the determinant of the matrix in (5.2.8).

Now if the determinant (5.2.8) is positive with $n = 1$, then it must be positive for all $n > 1$. So henceforth, we concentrate on the change of stability and associated bifurcation of steady-state solutions in the first eigenmode.

We assume

$$M(D\pi^2 + 1) < 1/4 \tag{5.2.9}$$

so that there exist two critical values for u_0 where (5.2.8) is zero. They are given by

$$u_{0\pm} = \frac{(1 - 2M(D\pi^2 + 1) \pm [1 - 4M(D\pi^2 + 1)]^{1/2})}{[2M(D\pi^2 + 1)]}. \tag{5.2.10}$$

By the above analysis, if $u_0 \in [0, u_{0-}) \cup (u_{0+}\infty)$, the solution (5.2.6) is locally stable while, for $u_0 \in (u_{0-}, u_{0+})$, the steady state is unstable in at least one eigenmode.

Now we shall seek solutions of the steady-state problem

$$\begin{aligned} & u_{xx} - (uc_x)_x = 0 \\ & Dc_{xx} - c + g(u) = 0 \qquad\qquad x \in (0, 1) \\ & c_x = u_x = 0 \;\text{ at }\; x = 0, 1 \end{aligned} \tag{5.2.11}$$

where $N(u)$ is close to $u_{0\pm}$, given by (5.2.10), and $h(u) \neq 0$. Since zero is a simple eigenvalue of the linearization at the steady state $(u_{0\pm}, g(u_{0\pm}))$, we are assured (see [8] and Chapter 2) that bifurcation takes place and we shall graph the resulting bifurcation curves in the (N, h)-plane.

Assume (5.2.9) holds so that both u_{0+} and u_{0-} in (5.2.10) exist. Then we claim there is a continuous branch of monotonically increasing steady-state solutions bifurcating from each of the rest states $(u_{0\pm}, g(u_{0\pm}))$ such that:

(i) on the branch through $(u_{0+}, g(u_{0+}))$

$$N(u) > N(u_{0+}) = u_{0+} :$$

(ii) on the branch through $(u_{0-}, g(u_{0-}))$

$$N(u) > N(u_{0-}), \text{ if } 0 < Q,$$

$$N(u) < N(u_{0-}), \text{ if } 0 > Q,$$

where $Q = Q(u_{0-})$, and $Q(u_0)$ is given by

$$\begin{aligned} Q(u_0) = &\frac{g_{uuu}(u_0)}{g_{uu}(u_0)} \\ &+ \frac{[u_0 g_{uu}(u_0) + (4\pi^2 D + 1)M]}{[u_0(M(4\pi^2 D + 1) - u_0 g_u(u_0))]}. \end{aligned} \tag{5.2.12}$$

To prove this, we introduce a small parameter ε^2 and seek solutions u so that $N(u) = u_0 + \varepsilon$, and $h(u) \neq 0$, where $u_0 = u_{0+}$ or u_{0-}.

Next we expand u and c in powers of $\varepsilon^{1/2}$. We obtain the result by applying a bifurcation analysis similar to that in section 1.5. This process is equivalent to the one introduced in the study of bifurcation in Chapter 2. Alternatively, this may be regarded as a kind of centre manifold approach (see Chapter 2 again and [7]).

Having obtained the local behaviour of the bifurcation curves near $u_{0\pm}$, we next investigate their global behaviour via a combination of analytical and numerical methods. Here we consider the steady state equations

$$\begin{aligned} &Mu_{xx} - (uc_x)_x = 0 \\ &Dc_{xx} - c + g(u) = 0 \end{aligned} \qquad x \in (0, 1) \tag{5.2.13}$$

$$0 = c_x = u_x \quad \text{at} \quad x = 0 \quad \text{and} \quad 1, \tag{5.2.14}$$

where g is $u/(1 + u)$, as before.

Using the boundary condition (5.2.14), the first of (5.2.13) integrates to give

$$u(x) = \lambda e^{c(x)/M} \tag{5.2.15}$$

where λ is any constant (positive since we require $u \geq 0$ everywhere). Substituting (5.2.15) into the second of (5.2.13), we obtain the problem

$$Dc_{xx} + G(c, \lambda) = 0 \tag{5.2.16}$$

where

$$G(c, \lambda) = g(\lambda e^{c/M}) - c. \tag{5.2.17}$$

Equation (5.2.16) is equivalent to the phase-space equations

$$\begin{aligned} c_x &= p \\ p_x &= -\,G(c, \lambda)/D \end{aligned} \tag{5.2.18}$$

and the boundary condition (5.2.14) reduces to

$$p = 0 \quad \text{at} \quad x = 0, 1. \tag{5.2.19}$$

The equilibria for (5.2.18) are given by $c = c^*$, $p = 0$ where $G(c^*, \lambda) = 0$. Since $G(0, \lambda) = g(\lambda) > 0$ and $G(c, \lambda) < 0$ for large c, $G(c, \lambda)$ must have an odd number of zeros (counting multiplicities).

It turns out that there exist constants $0 < \lambda_1 < \lambda_2$ such that G has exactly three simple zeros when $\lambda \in (\lambda_1, \lambda_2)$ and exactly one zero in $[0, \infty)/[\lambda_1, \lambda_2]$. λ_1 and λ_2 depend solely on M. λ_1 and λ_2 are determined from the requirement that $G(c, \lambda)$ has a double root. The actual algebraic expressions, which can easily be obtained, are not required here.

Now for $\lambda \notin [\lambda_1, \lambda_2]$, the phase portrait for (5.2.18) contains a simple saddle point and it is impossible to have a trajectory starting out from and finishing on the line $p = 0$. For $\lambda \in (\lambda_1, \lambda_2)$, the phase portrait for (5.2.18) contains two saddles at $c = c_1, c_3$ and a centre at $c = c_2$ (see Figure 5.1). Here, it is possible to have trajectories such as AB in Figure 5.1 which start out from and finish on the line $p = 0$. Such orbit segments will correspond to solutions of (5.2.13) and (5.2.14) if the *time* taken up in x between A and B is one.

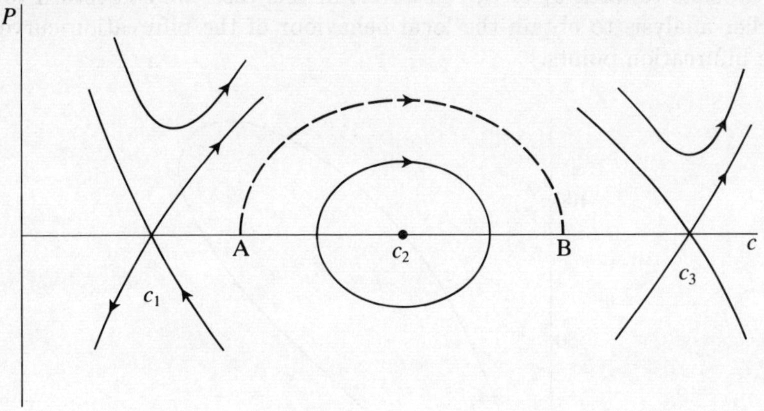

Figure 5.1: Typical phase plane for (5.2.18)

For $\lambda \in (\lambda_1, \lambda_2)$ fixed if we vary the position of A between the roots c_1 and c_2 we may define the time-map $T(A)$ which represents the time taken for the orbit segment through A to arrive at the corresponding point B on the line $p = 0$. In fact, $T(A)$ may be represented via an elliptic integral and is certainly continuously defined in an interval to the left of c_2 (see section 1.5). Moreover, $T(A)$ increases as A moves from c_2 towards c_1 and tends to infinity either as A approaches c_1 or as the corresponding point B approaches c_3. For these and related matters, see [70].

Thus for each λ fixed, there is at most one trajectory of the form AB in Figure 5.1 for which $T(A) = 1$. Note that such a trajectory yields monotone increasing solutions $(u(x), c(x))$ of (5.2.13), (5.2.14). The symmetry in the phase plane implies that the orbit segment BA in the lower half plane provides the corresponding monotone decreasing steady state which is inferred via the symmetry of the original problem.

By fixing $\lambda \in (\lambda_1, \lambda_2)$, (5.2.18) may be solved numerically and we may use the time-map to seek solutions satisfying (5.2.19). Using the continuous dependence upon λ, we may graph the resulting solution branch in the (N, h)-plane.

The advantage of this approach is that we obtain equilibrium solutions regardless of their stability as solutions of the time-dependent problem (5.2.4)-(5.2.5).

The results are summarized in Figures 5.2-5.3, where we show the bifurcation curves for monotone increasing steady-state solutions when M and D are fixed appropriately.

Finally, we point out that the numerical procedure becomes inefficient close to the bifurcation points ($N = u_{0\pm}$) since here the middle root c_2

of G tends towards c_1 or c_3. However, in this case, we can return to our earlier analysis to obtain the local behaviour of the bifurcation curves at the bifurcation points.

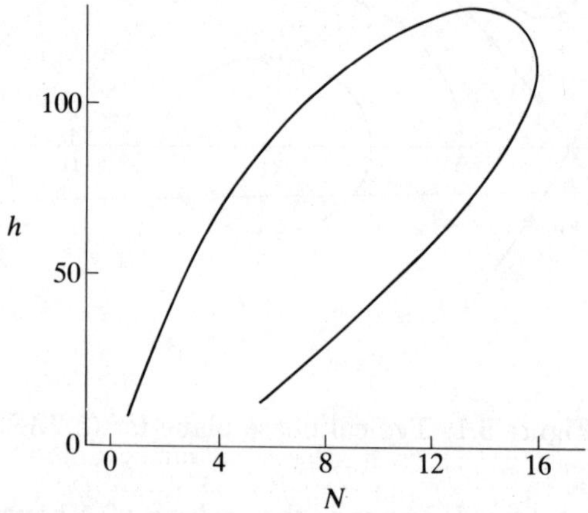

Figure 5.2: Numerically obtained bifurcation diagram, for $M = 0.1$, $D = 0.05$ ($u_{0+} = 4.47$, $u_{0-} = 0.224$).

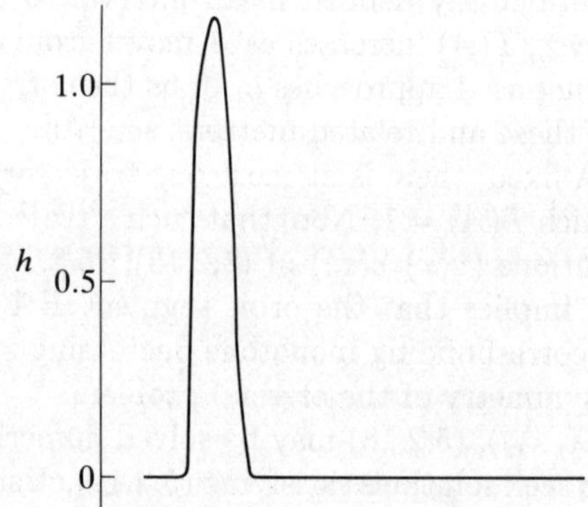

Figure 5.3: Numerically obtained bifurcation diagram, for $M = 0.1$, $D = 0.15$ ($u_{0+} = 1.19$, $u_{0-} = 0.838$).

We may solve (5.2.2) numerically to obtain stable steady state solutions; see [28]. The equilibrium shown in Figure 5.4 displays characteristic peaks, and can be obtained by extending our monotone increasing solutions periodically.

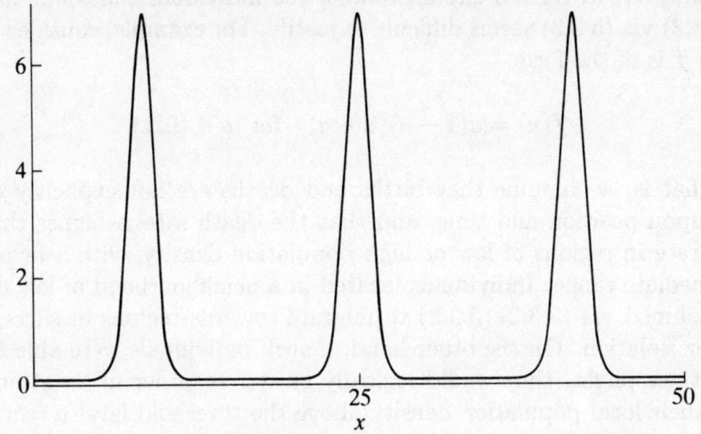

Figure 5.4: Spatial pattern in cell density obtained by solving (5.2.3), with no-flux boundary conditions, and $M = 0.125$, $D = 1$; from [28].

5.3 Aggregation in population biology

In many models used to study the spatial dispersion of biological populations, theory has been based upon equations of the form

$$u_t = \Delta u + f(u, \mathbf{x}, t) \tag{5.3.1}$$

$$u_t = \Delta(\phi(u)) + f(u, \mathbf{x}, t). \tag{5.3.2}$$

Here, $u(\mathbf{x}, t)$ denotes the population density at location $\mathbf{x}$ and time t; $f(u, \mathbf{x}, t)$ represents the rate at which individuals are supplied to the population directly at $\mathbf{x}$, at time t, due to births and deaths; ϕ in (5.3.2) is a function satisfying

$$\begin{aligned} \phi(0) &= 0 \\ \phi_u(u) &> 0 \quad \text{for} \quad u > 0. \end{aligned} \tag{5.3.3}$$

In deriving the reaction-diffusion equation (5.3.1), it is assumed that the dispersal of individuals is due to random motion (see Chapter 1), whilst the derivation of (5.3.2) rests on the assumption that individuals disperse to avoid crowding (see section 5.4).

However, in certain circumstances, the individual behaviour imposed in (5.3.2) via (5.3.3) seems difficult to justify. For example, consider (5.3.2) where f is of the form

$$f(u) = u(1-u)(u-a) \quad \text{for} \quad a \in (0,1). \tag{5.3.4}$$

That is, we assume that births and deaths are not explicitly dependent upon position and time, and that the death rate is higher than the birth rate in regions of low or high population density, with a favourable intermediate range. Individuals located in a neighbourhood of low density are assumed via (5.3.2),(5.3.3) to migrate towards regions of successively greater isolation. On the other hand, if such individuals were able to perceive their plight, they would logically cluster together in an attempt to raise their local population density above the threshold level $u = a$, when the expected birth rate would begin to outweigh the expected death rate.

The above scenario and its remedy would only be relevant where individuals are able to utilize information about their local habitat to trigger a natural response in their behaviour. However, phenomena such as swarming, herding, and more local grouping or clustering of individuals are relatively common forms of behaviour and provide a local environment in which individuals are able to hunt, mate, and defend themselves more effectively. In short, by forming local aggregations, individuals may increase the chances of survival for themselves and their offspring [68].

Clearly, the models based on (5.3.1) and (5.3.2) contain no mechanism by which individuals aggregate. Moreover, in terms of the individual's response to the environment, individual births and deaths probably occur on a much longer time scale; so, what is required is a model in which individuals are able to form aggregates in response to the *threat* of predation, starvation, or other externally controlled factors. We follow [27].

Let $u(\mathbf{x}, t)$ denote the population density at $\mathbf{x}$, at time t. Then the population balance equation is

$$u_t = -\nabla.(u\mathbf{v}) + f(u, \mathbf{x}, t). \tag{5.3.5}$$

Here, $\mathbf{v}$ denotes the average velocity of individuals at $\mathbf{x}$, at time t, and f is as in (5.3.1). More specifically, we define $E(u, \mathbf{x}, t)$ to be the projected net rate of reproduction per individual at $\mathbf{x}$, at time t, within a local population density u. That is

$$E = \text{average birth rate} - \text{average death rate}$$

where the average is taken over all individuals at $\mathbf{x}$, at time t, while $u(\mathbf{x}, t) = u$.

Clearly

$$f = uE(u, \mathbf{x}, t) \tag{5.3.6}$$

so that E defines f or vice versa.

We assume that individuals execute random walks, biased by some ideal or optimal velocity, $\mathbf{w}$ (see Box A on the Fokker-Planck equation). An individual with velocity $\mathbf{w}$ disperses deterministically so as to increase its expected rate of reproduction so that, by taking random walks biased by $\mathbf{w}$, the population is, on average, dispersing in the ideal direction. We have

$$\mathbf{v} = -\delta\frac{\nabla u}{u} + \mathbf{w} \tag{5.3.7}$$

where δ is some nonnegative constant representing the random dispersion of individuals (see section 1.2). In order to define $\mathbf{w}$, we first observe that $\mathbf{w}$ should be in the direction of increasing $E(\mathbf{x}, t)$, say

$$\mathbf{w} \approx \text{ constant } \times \nabla E(u(\mathbf{x}, t), \mathbf{x}, t).$$

We modify this slightly by setting

$$\mathbf{w} = \int k(\mathbf{x}, \mathbf{y})\nabla E(\mathbf{y}, t)\, dy \tag{5.3.8}$$

for some suitable nonnegative kernel $k(\mathbf{x}, \mathbf{y})$. In this way, $\mathbf{w}$ is a **local average** of ∇E. For simplicity of analysis, a particularly attractive choice is to define k to be the Green function for the operator $(-\varepsilon\Delta + I)$, defined on the domain Ω, representing the habitat of the population. In this way, we may also impose boundary conditions upon $\mathbf{w}$, making sure that individuals do not attempt to leave or enter Ω deterministically.

We have

$$-\varepsilon\Delta\mathbf{w} + \mathbf{w} = \lambda\nabla E \tag{5.3.9}$$

where $\varepsilon > 0$ is a small constant, and λ is some scaling, nonnegative constant. The effect of the first term in (5.3.9) is simply to smooth out any sharp local variations in ∇E so that $\mathbf{w}$ is a local average of the optimal velocity $\lambda\nabla E$. (By taking Fourier transforms, for example, one may easily see that $\mathbf{w}$ is less responsive to components of ∇E giving rise to high wave numbers.)

Putting (5.3.5) and (5.3.9) together, we have

$$\begin{aligned} u_t =& \delta\Delta u - \nabla(u\mathbf{w}) + uE(u, \mathbf{x}, t) \\ -& \varepsilon\Delta\mathbf{w} + \mathbf{w} = \lambda\nabla E(u, \mathbf{x}, t) \end{aligned} \tag{5.3.10}$$

which must hold for $t > 0$ and $\mathbf{x}$ in some domain Ω in $\mathbf{R}^n$. At the boundaries of Ω, we shall impose no-flux boundary conditions

$$\begin{aligned} \mathbf{n}.\nabla u &= 0 \\ \mathbf{n}.\mathbf{w} &= 0 \end{aligned} \quad \mathbf{x} \in \delta\Omega,\ t \geq 0, \tag{5.3.11}$$

which ensures that no individuals may leave or enter the domain Ω at the boundaries.

Rescaling

$$\tilde{\mathbf{w}} = \mathbf{w}/\lambda, \quad \tilde{t} = t\lambda, \quad \tilde{\delta} = \delta/\lambda, \quad r = 1/\lambda,$$

then dropping the tildes, (5.3.10) becomes

$$\begin{aligned} u_t &= \delta\Delta u - \nabla(u\mathbf{w}) + ruE(u, \mathbf{x}, t) \\ -\varepsilon\Delta\mathbf{w} + \mathbf{w} &= \nabla E(u, \mathbf{x}, t), \end{aligned} \quad \mathbf{x} \in \Omega,\ t \geq 0. \tag{5.3.12}$$

In what follows, we shall concentrate on models in homogeneous environments, so that E in (5.3.12) will not depend explicitly on $\mathbf{x}$ or t. We have

$$\begin{aligned} u_t &= \delta\Delta u - \nabla.(u\mathbf{w}) + ruE(u) \\ -\varepsilon\Delta\mathbf{w} + \mathbf{w} &= E_u(u)\nabla u, \end{aligned} \quad \mathbf{x} \in \Omega,\ t \geq 0. \tag{5.3.13}$$

For example, if $E(u) = (1-u)(u-a)$ for some constant $a \in (0,1)$, we have the bistable growth rate of (5.3.4).

Equations (5.3.13),(5.3.11) may be solved numerically in the domain $\Omega = [0, 10] \subset \mathbf{R}$, with $r = 1.0, \delta = 0.1, \varepsilon = 0.01, a = 0.20$, and choosing random initial values $u(x, 0)$ in [0.17,0.19].

It is not difficult to show via comparison principle arguments (see Chapter 1) that the corresponding boundary-value problems for (5.3.1) or (5.3.2) yield solutions which decay to zero.

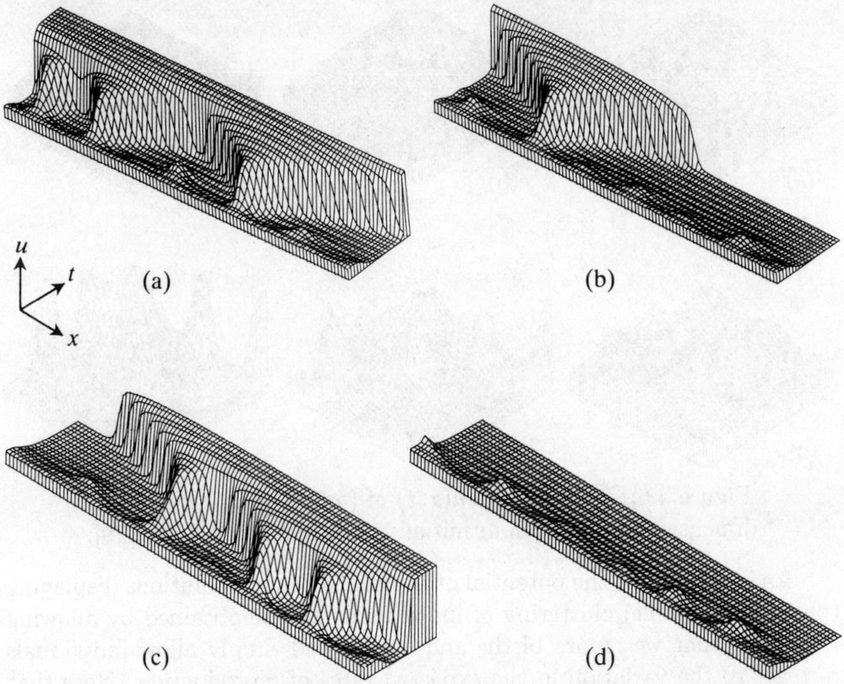

Figure 5.5(a)-(d): The solution $u(x,t)$ of (5.3.13),(5.3.11). Four examples with random initial values $u(x,0)$ in [0.17,0.19]. Results are plotted for $0 \leq t \leq 50$.

Figure 5.5(a)-(d) depicts solutions for different random initial values. The first three show that local clusters form, some of which become dense enough to raise the solution above threshold $(u - a)$ and enable the population to approach the saturated steady state $(u = 1)$.

In Figure 5.5(d), despite the formation of local clusters, the population is not able to react fast enough and the solution finally decays away to zero.

Clearly, the examples depicted in Figure 5.5(a)-(d) contain a rich structure of time-dependent solutions. In particular, we can see that our simple model of aggregative phenomena yields solutions which display the behavioural response anticipated in our discussion in the introduction.

Figure 5.6 shows a numerical solution of (5.3.13) (5.3.11) on a square domain in two dimensions. Solutions are plotted at equal time intervals; the parameters and initial conditions are as for Figure 5.5.

Before moving on to consider some multi-species interactions, we pause to make some general remarks concerning our model (5.3.13), and some of its special cases.

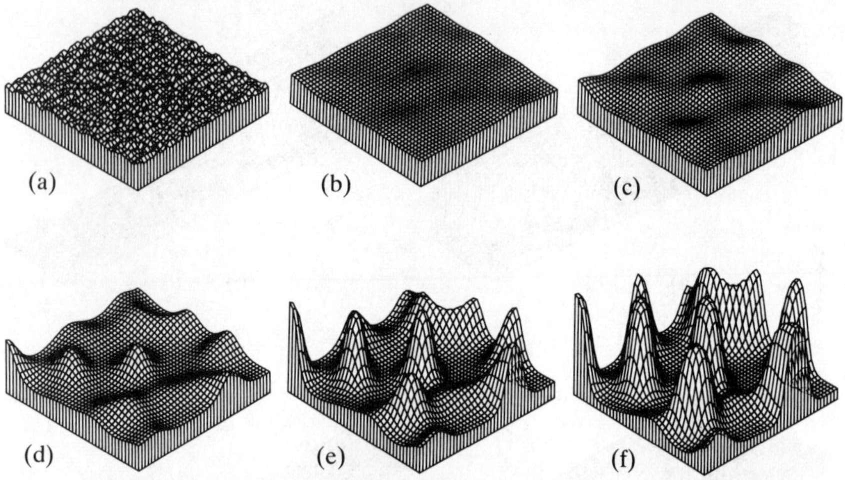

Figure 5.6: The solution $u(x,t)$ of (5.3.13),(5.3.11) in two dimensions with random initial values $u(x,0)$ in [0.17,0.19].

An indication of the potential of the model to yield solutions displaying (possibly transient) clustering of individuals may be obtained by allowing $r \to 0$, so that we ignore births and deaths and simply allow individuals to react to the variation in the expected rates of reproduction. Note that $r \approx 0$ is equivalent to assuming that births and deaths take place on a much longer time scale than the dispersal of individuals. We have

$$\begin{aligned} u_t = \delta\Delta u - \nabla(u\mathbf{w}), \quad & x \in \Omega, \ t \geq 0. \\ -\varepsilon\Delta\mathbf{w} + \mathbf{w} = E'(u)\nabla u, & \\ \mathbf{n}.\nabla u = \mathbf{n}.\mathbf{w} = 0, \quad & x \in \delta\Omega, \ t \geq 0. \end{aligned} \tag{5.3.14}$$

Clearly, $d/dt \int_\Omega u\,dx = 0$, so we may linearize (5.3.14) about any steady-state $u \equiv u_0 > 0, \mathbf{w} \equiv 0$, where u_0 is arbitrary. Note that when r is small, steady-state solutions of (5.3.14) may be thought of as pseudo steady-states for (5.3.13).

Let $0 = \lambda_0 \leq \lambda_1 \leq \lambda_2 ...$ denote the eigenvalues of $-\Delta$ on Ω subject to Neumann (no-flux) boundary conditions. Then the steady-state solution is unstable in the $n(\geq 1)$th eigenmode if and only if

$$\delta(\varepsilon\lambda_n + 1) < u_0 E_u(u_0).$$

Thus, if $E_u(u_0) > 0$ and δ, ε are small, random perturbations of $u \equiv u_0$ will produce spatially structured solutions. When E is as in the example in Figure 5.5 and $u_0 \approx a$, this is certainly the case.

The most obvious simplification of (5.3.13) is to allow $\varepsilon = 0$ so that we may substitute for $\mathbf{w}$ and obtain the single equation

$$u_t = \nabla.\{(\delta - uE_u(u))\nabla u\} + ruE(u). \tag{5.3.15}$$

This is a useful equation when u is such that $\delta - uE_u(u) \geq 0$, but is not wellposed if the initial values enter a range where $uE_u(u) \leq \delta$. The problem is that such examples would generally yield solutions which develop sharp discontinuities (e.g. shock waves) even when the initial values were smooth. This makes both analytical and numerical work much more involved and undermines the essential simplicity of the model. Heuristically, when $\varepsilon = 0$, the deterministic velocity $\mathbf{w}$, defined in (5.3.9), becomes too sensitive to extremely local fluctuations in E. As a result, the population distribution may become discontinuous when neighbouring individuals decide to disperse in opposite directions. For $\delta > 0, \varepsilon > 0$, we can show that solutions for (5.3.13) exist locally, provided $u(0, x)$ is in $H_1(\Omega)$, using the techniques in chapter 1.

In some cases, such as so-called logistic growth when $E(u) = 1 - u$, say, the limits $\varepsilon \to 0$ and $\delta \to 0$ are justified, and (5.3.13) takes the qualitative form of (5.3.2). In this example though, since $E_u < 0$, the individuals dispersing so as to maximize E would seek isolation, and there is clearly no mechanism capable of producing aggregation of individuals.

Next, we consider two competing species with densities u_1, u_2, present in some domain Ω. We assume for simplicity that the individual net reproduction rates are given by

$$\begin{aligned} E_1 &= A - au_1 - bu_2 \\ E_2 &= B - a^*u_1 - b^*u_2 \end{aligned} \tag{5.3.16}$$

for species 1 and 2 respectively. Here, A, B, a, a^*, b, b^* are positive constants. Notice that the ordinary differential equations

$$\begin{aligned} u_{1t} &= u_1 E_1 \\ u_{2t} &= u_2 E_2 \end{aligned}$$

are the usual Lotka-Volterra competition equations.

Initially, suppose that

$$ab^* < a^*b \tag{5.3.17}$$

which means that interspecific competition is stronger than intraspecific competition.

Our model for the dispersal of individuals of this two species competition is given by utilizing a scaled version of (5.3.10) for each species. That is

$$\begin{aligned} u_{1t} &= \delta_1 \Delta u_1 - \nabla.(u_1 \mathbf{w}_1) + r u_1 E_1(u_1, u_2) \\ u_{2t} &= \delta_2 \Delta u_2 - \mu \nabla.(u_2 \mathbf{w}_2) + r u_2 E_2(u_1, u_2) \\ -\varepsilon_1 \Delta \mathbf{w}_1 + \mathbf{w}_1 &= \nabla E_1 = -a \nabla u_1 - b \nabla u_2 \\ -\varepsilon_2 \Delta \mathbf{w}_2 + \mathbf{w}_2 &= \nabla E_2 = -a \nabla u_1 - b^* \nabla u_2 \end{aligned} \tag{5.3.18}$$

for $x \in \Omega, t \geq 0$.

Again, we shall impose no-flux boundary conditions

$$\mathbf{n}.\nabla u_i = \mathbf{n}.\mathbf{w}_i = 0, \quad i = 1, 2 \text{ for } \mathbf{x} \in \partial\Omega,\ t \geq 0. \tag{5.3.19}$$

Guided by our experience with the one-species model, we can check whether the nonlinear dispersal mechanism is likely to induce clustering-type behaviour by examining (5.3.18)-(5.3.19) in the limit $r \to 0$. Again, this is effectively assuming that dispersal events take place on a much shorter time scale than birth-death events. With $r = 0$, we see that

$$\begin{aligned} u_1 &\equiv u_{1,0}, \quad \mathbf{w}_1 \equiv 0 \\ u_2 &\equiv u_{2,0}, \quad \mathbf{w}_2 \equiv 0 \end{aligned} \tag{5.3.20}$$

is a steady-state solution for (5.3.18)-(5.3.19), for any constants $u_{1,0}, u_{2,0}$, nonnegative. Linearizing (5.3.18) about (5.3.20), we proceed as before and can show that the uniform steady-state (5.3.20) is unstable in the nth eigenmode ($n \geq 1$) if and only if

$$\begin{aligned} &\delta_1\delta_2 + \delta_1 \mu u_{2,0} b^* / (\varepsilon_2 \lambda_n + 1) + \delta_2 u_{1,0} a / (\varepsilon_1 \lambda_n + 1) \\ &\quad < \mu u_{1,0} u_{2,0} (a^* b - b^* a) / ((\varepsilon_1 \lambda_n + 1)(\varepsilon_2 \lambda_n + 1)) \end{aligned} \tag{5.3.21}$$

where λ_n are the same as before.

In view of (5.3.17), it is clear that, for each $n = 1, 2, ...$, we obtain a region in $(u_{1,0}, u_{2,0})$-space in which (5.3.21) holds (see Figure 5.7). Hence, in such regions, perturbations of (5.3.20) would induce spatially structured solutions.

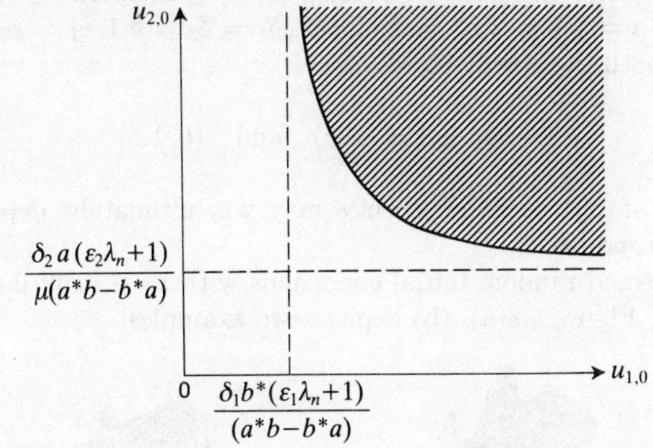

Figure 5.7: Schematic Figure of the region of instability in the nth eigenmode for spatially homogeneous solutions of (5.3.18),(5.3.19) (with $r = 0$).

Note that, if (5.3.17) does not hold, then (5.3.21) never holds for any $u_{1,0}, u_{2,0}$ positive. Hence, if intraspecific competition dominates interspecific competition, the uniform population density functions are always stable. This confirms the intuitive notion that, in this case, at least one species is more threatened by intraspecific competition and hence such individuals avoid aggregation.

Notice that, if a^* is allowed to become negative in (5.3.16) while a, b, b^* remain positive, we obtain a simple predator-prey model. Here, u_1 denotes the prey density. In this case, (5.3.17) can never hold and hence no spatial patterns develop when the uniform distributions are perturbed.

Returning to the original problem, (5.3.18)-(5.3.19) with $r > 0$, we expect that the initial-value problem will initially exhibit aggregative features as individuals initially seek to minimize the threat of interspecific competition. However, the birth-death population supply terms should allow this transient aggregative phase to give way to a more long-term regime during which individuals of one species are able gradually to deplete the individuals of the other species until ultimately, all individuals of the weaker species die off and the stronger species saturates.

As $t \to \infty$, we expect either $u_1 \to A/a$ and $u_2 \to 0$ everywhere, or $u_1 \to 0$ and $u_2 \to \Omega/b^*$ everywhere. Clearly, a stable steady-state for (5.3.18)-(5.3.19) must also be a stable rest point for the associated ordinary differential equations $u_{1t} = u_1 E_1, u_{2t} = u_2 E_2$; and the condition (5.3.17) ensures that a rest point with both u_1 and u_2 nonzero (if it exists) cannot be stable.

We solved (5.3.18) on the domain $[0,10] \subseteq \mathbf{R}$, with $r = 1, \mu = 1, A = 1, B = 1.5, a = 1, b = 2, a^* = 3, b^* = 1, \delta_1 = \delta_2 = 0.1, \varepsilon_1 = \varepsilon_2 = 0.025$. In this case, both the steady-states

$$(u_1, u_2) = (1, 0) \quad \text{and} \quad (0, 1.5)$$

are locally stable, so either species may win ultimately, depending upon the initial conditions.

We selected random initial conditions with $u_1 \in [0.39, 0.41]$ and $u_2 \in [0.29, 0.31]$. Figure 5.8(a), (b) depicts two examples.

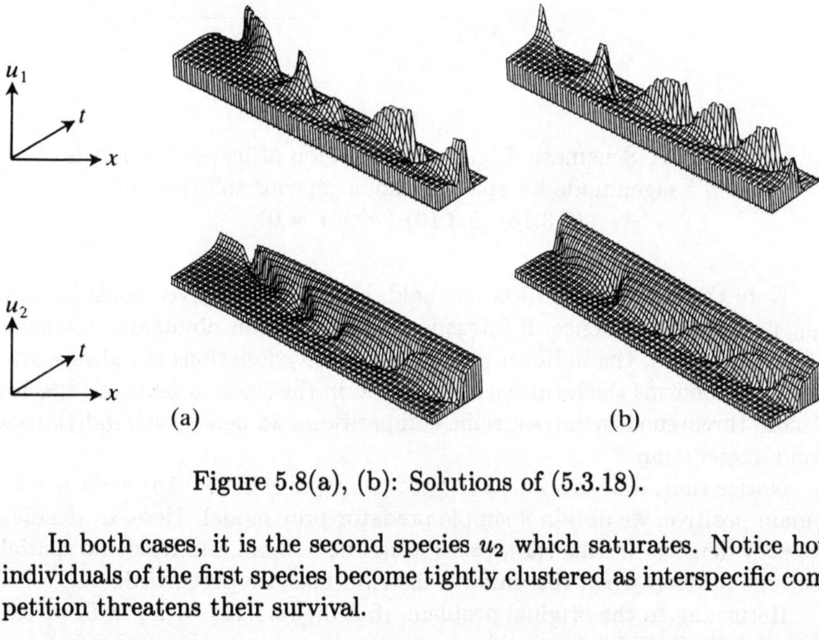

Figure 5.8(a), (b): Solutions of (5.3.18).

In both cases, it is the second species u_2 which saturates. Notice how individuals of the first species become tightly clustered as interspecific competition threatens their survival.

We have shown how spatial heterogeneity arises when births and deaths are assumed to occur on a much longer time scale than the dispersive behaviour of individuals.

In general, although there is much interest in dispersal amongst individuals, and its possible consequences, it is not straightforward to pick out real examples in which everyone would agree that the dispersal is dominantly nonrandom. However, models such as (5.3.13) have begun to address such situations, primarily to suggest which types of spatial patterns can possibly be due to deterministic dispersive behaviour, and which require external forcing via convection or resource variability, for example. These models ((5.3.13) etc.) do not support any real clustering phenomena though; and, given the often patchy nature of population distributions, there is a clear need for models which can predict the effects of deterministic dispersal in producing clustering-type behaviour.

It is easy to think of human populations forming small nucleated settlements which grow as the population saturates locally. Villages become towns, which grow together, forming large conurbations. Such behaviour is exactly that of Figure 5.5(a)-(d). In view of the types of models discussed here, it would be advantageous if data was available which measured not only population distributions but also the population flux. Any dependence of one on the other via nonlinear mechanisms, such as those we propose, could then be sought.

Next, we discuss the travelling wave solutions of the simplest aggregation model, following [26]. We consider the single-species model with bistable growth rate given by (5.3.4) for $x \in \mathbf{R}$. Introducing the travelling variable $\xi = x + ct$, we seek a solution of (5.3.13) so that, writing $u = u(\xi), w = w(\xi)$, we obtain

$$\begin{aligned} cu_\xi &= [\delta u_\xi - uw]_\xi + uE(u) \\ -\varepsilon w_{\xi\xi} + w &= E(u)_\xi = E_u(u)u_\xi \end{aligned} \tag{5.3.22}$$

where $E_u = dE/du$; $E(u) = (1-u)(u-a)$ for some constant $a \in (0,1)$ (recall that the constants δ and ε are positive with $1 \gg \varepsilon > 0$).

We seek a solution $(c, u(\xi), w(\xi))$ of (5.3.22) such that

$$\lim_{\xi \to \infty} (u, w) = (1, 0)$$
$$\lim_{\xi \to -\infty} (u, w) = (0, 0).$$

As a first step, set $\varepsilon = 0$, so that (5.3.22) becomes

$$cu_\xi = [(\delta - uE_u)u_\xi]_\xi + uE(u). \tag{5.3.23}$$

For any constant c, solutions of (5.3.23) may be sketched as trajectories in the phase plane. Note, however, that, if $\delta = uE_u(u)$ for some $u^* \in (0,1)$, then, as $u \to u^*$ along trajectories, u_ξ must become unbounded.

We split our analysis into two cases:
(i) $\delta > uE_u(u) = u((1+a) - 2u)$ for $u \in [0,1]$;
(ii) $\delta = uE_u(u) = u((1+a) - 2u)$ at u_1 and u_2, say, where $0 < u_1 \le u_2 < 1$, that is, u_1 and u_2 are simple zeros of $\delta - uE_u$ and $\delta - uE_u$ is positive in $(0, u_1)$ and $(u_2, 1)$.

Case (i).

In this case, there exists a unique value c such that (5.3.23) possesses a solution connecting the rest states $u = 0$, $u = 1$. To see this, we set

$$\xi = \int_0^y (\delta - v(y)E_v(v(y)))\, dy$$

and

$$u(\xi) = v(y)$$

where

$$cv_y = v_{yy} + (\delta - vE_v(v))vE(v). \tag{5.3.24}$$

It is a simple matter to show that $u(\xi)$ is the required solution if and only if $v(y)$ satisfies (5.3.24) together with

$$\lim_{y\to\infty} v = 1$$

$$\lim_{y\to-\infty} v = 0$$

and $\delta > vE_v(V)$ for all $y \in \mathbf{R}$.

The equation (5.3.24) is of the general form previously considered in section 1.5. Since $\delta > vE_v(v)$, and $vE(v)$ is qualitatively of *cubic* form, there exists a monotone increasing solution $v(y)$, for a unique wave speed c. Moreover, the sign of c is determined by

$$\operatorname{sgn}(c) = \operatorname{sgn}\{\int_0^1 (\delta - vE_v(v))vE(v)dv\}$$

which, in this case, is positive if and only if

$$0 < \delta(1-2a)/12 + (1-4a+5a^2)/60.$$

Having obtained a front solution $(u_0(\xi), c_0)$, say, we expect that, for ε small, (5.3.22) possesses an asymptotic solution of the form

$$u = u_0(\xi) + 0(\varepsilon), \quad w = u_{0\xi}E_u(u_0) + 0(\varepsilon)$$

for some wave speed $c = c_0 + 0(\varepsilon)$.

Case (ii).

Consider the phase-plane representation for (5.3.23),

$$\begin{aligned} u_\xi &= p \\ (\delta - uE_u)p_\xi &= p^2(E_u + uE_{uu}) + cp - uE(u). \end{aligned}$$

Linearizing about the rest points $(0,0)$ and $(1,0)$, we see that both are saddle points for any value of c. However, the unstable manifold of $(0,0)$ leading into the first quadrant cannot reach the line $u = u_1$, nor can the stable manifold of $(1,0)$ ever be traced back to the line $u = u_2$. Thus, no continuous trajectory may ever connect the two saddle points.

We can, however, construct weak solutions of (5.3.23) which allow a single jump discontinuity in u and u_ξ. Such solutions correspond to travelling shock solutions of the original partial differential equation (5.3.4) (with $\varepsilon = 0$).

We proceed as follows. Suppose for $\xi < 0$, u satisfies (5.3.23) and is such that $\lim_{\xi\to-\infty}(u, u_\xi) = (0,0)$. For $\xi > 0$, we suppose that u satisfies (5.3.23) together with $\lim_{\xi\to\infty}(u, u_\xi) = (1,0)$. Clearly, each *half* solution is well defined up to a translation in ξ for all $c \in \mathbf{R}$. At $\xi = 0$, we apply the usual shock condition (continuity of flux)

$$cu|_-^+ = (u_\xi(\delta - uE(u)))|_-^+. \tag{5.3.25}$$

Here

$$u|_-^+ = \lim_{\xi\to 0,\xi>0} u(\xi) - \lim_{\xi\to 0,\xi<0} u(\xi) = u^+ - u^-.$$

We must choose $u^+(\xi)$, $u^-(\xi)$, and c so as to satisfy (5.3.25) together with the condition that each half solution lies on the appropriate manifold in the phase space. By introducing the variable y defined in (5.3.24), it is possible to show that, if $u^\pm$ are held fixed as c varies, u_ξ^+ and u_ξ^- may be considered as functions of c alone and satisfy

$$\begin{aligned}
u_\xi^+ &\to \infty \;\text{ as }\; c \to -\infty,\\
u_\xi^+ &\to 0 \;\text{ as }\; c \to c_1 > 0,\\
u_\xi^- &\to \infty \;\text{ as }\; c \to \infty,\\
u_\xi^- &\to 0 \;\text{ as }\; c \to -c_2 < 0,
\end{aligned}$$

where c_1 and c_2 are some positive constants. Moreover, u_ξ^+ is monotone decreasing, whilst u_ξ^- is monotone increasing.

The half solutions defined above will be utilized as candidates for outer solutions in our matched asymptotic expansion, so we shall denote them by $u_{Out}^\pm$, etc..

In a neighbourhood of $\xi = 0$, both u_{Out} and $u_{Out\ \xi}$ suffer possible jump discontinuities. Returning to (5.3.22), we stretch the independent variable by setting $\xi = \varepsilon\zeta$ and rescale w by setting $w = p\varepsilon^{-1/2}$. To order ε^0, (5.3.22) becomes

$$(\delta u_\zeta - up)_\zeta = 0 \tag{5.3.26}$$

$$-p_{\zeta\zeta} + p = uE_u(u). \tag{5.3.27}$$

Here we must seek a solution so that

$$\begin{aligned}
(u, p, p_\zeta) &\to (u_{Out}^+(0), 0, 0) \;\text{ as }\; \zeta \to \infty\\
(u, p, p_\zeta) &\to (u_{Out}^-(0), 0, 0) \;\text{ as }\; \zeta \to -\infty.
\end{aligned} \tag{5.3.28}$$

If this is done, the inner solution is matched to the outer solution up to $O(\varepsilon^0)$, so the matched solution is simply the first term in an asymptotic expansion for the travelling solution of (5.3.4) (cf. case (i) above). Now, (5.3.26) and (5.3.27) imply

$$\begin{aligned} \delta u_\zeta &= up \\ p_\zeta &= s \\ s_\zeta &= p - u_\zeta E_u(u). \end{aligned} \tag{5.3.29}$$

Therefore

$$s_\zeta = (p/\delta)[\delta - E_u(u)] = u_\zeta[\delta/u - E_u(u)]$$

so that

$$s = \delta \ \ln u - E(u) + k \tag{5.3.30}$$

where k is some constant. Now using (5.3.30) in the second of (5.3.29) and multiplying by p, we obtain

$$pp_\zeta = p(\delta \ \ln u - E(u) + k) = \delta u_\zeta/u(\delta \ \ln u - E(u) + k).$$

Thus

$$p^2/2\delta = \int_{u_{Out}^-(0)}^{u} [\delta \ (\ln u)/v - E(v)/v + k/v] \, dv. \tag{5.3.31}$$

Here, we have used $\lim_{\zeta\to-\infty}(u,p) = (u_{Out}^-(0))$. Imposing $\lim_{\zeta\to\infty}(u,p) = (u_{Out}^+(0))$, (5.3.31) implies that

$$\int_{u_{Out}^-(0)}^{u_{Out}^+(0)} [(\delta \ \ln u - E(v) + k)/v] dv = 0. \tag{5.3.32}$$

Also, from (5.3.30), we see that

$$0 = \delta \ \ln u_{Out}^{\pm}(0) - E(u_{Out}^{\pm}(0)) + k.$$

By hypothesis, $d/du(\delta \ \ln u - E(u))$ changes sign at u_1 and u_2 and $\delta \ \ln u - E(u)$ has a maximum at u_1 and a minimum at u_2. Thus, for each $k \in (\delta \ \ln u_2 - E(u_2), \delta \ \ln u_1 - E(u_1))$, $\delta \ \ln u - E(u)$ has three roots and we are forced to choose $u_{Out}^-(0)$ to be the first, smallest root and $u_{Out}^+(0)$ to be the third, largest root, so that the integrand in (5.3.32) changes sign. Moreover, if $k = \delta \ \ln u_2 - E(u_2)$, the left hand side of (5.3.32) is positive, whilst if $k = \delta \ \ln u_1 - E(u_1)$, it is negative. Thus, there exists a value for k and hence constants $u_{\rm Out}^{\pm}(0)$ such that (5.3.30) and (5.3.32) hold. Hence, (5.3.29) possesses a solution as required. It is relatively straightforward to show that the left-hand side of (5.3.32) is monotone decreasing as a function of k. Hence, the transitional, inner solution, together with its asymptotes $u^{\pm}_{\text{Out}}(0)$, is determined uniquely.

Returning to the outer solution, the values of $u^{\pm}_{Out}(0)$ are now fixed so that the derivatives $u^{\pm}_{Out\ \xi}(0)$ can be regarded as functions of c. The condition (5.3.26), together with the monotonic dependence of $u^{\pm}_{Out\ \xi}(0)$ upon c, yields a unique solution for c and hence for the outer solution $u_{Out}(\xi)$.

Our asymptotic solution yields the value of c to $O(\varepsilon^0)$. In practice, as ε varies, c will vary. In order to estimate this behaviour for ε small, it would be necessary to consider the next term in our expansions for u and w.

If

$$\begin{aligned} u &= u^* + \varepsilon u_1 + O(\varepsilon^2) \\ w &= w^* + \varepsilon w_1 + O(\varepsilon^2) \\ c &= c^* + \varepsilon c_1 + O(\varepsilon^2), \end{aligned}$$

where (u^*, w^*, c^*) represents the matched asymptotic solution obtained above, then u_1, w_1, and c_1 must satisfy an equation of the form

$$L(u_1, w_1) = h(c_1), \tag{5.3.33}$$

where L is a linear differential operator with domain in $L^2(\mathbf{R}) \times L^2(\mathbf{R})$, depending upon u^*, w^*, and c^*, and $h \in L^2(\mathbf{R}) \times L^2(\mathbf{R})$ is a function of c_1, together with u^*, w^*, c^*.

Since travelling waves are translation invariant, symmetry implies

$$L(u^*_\xi, w^*_\xi) = 0.$$

Thus, c_1 must be chosen so that $h \in$ range (L) and hence a solution (u_1, w_1) orthogonal to (u^*_ξ, w^*_ξ) exists. Thus, c_1 is determined by applying the Fredholm alternative to (5.3.33) (see Box G).

5.4 Nonlinear diffusion

The classical theory of Fickian diffusion is based on individual particles executing a certain type of random walk; see section 1.2. Whilst being an appropriate model for many diverse transport processes, it is sometimes unrepresentative, particularly at extremely low or high densities. For example, consider the one-dimensional diffusion equation

$$u_t = u_{xx}, \qquad x \in \mathbf{R}, \quad t \geq 0,$$

together with the initial values

$$u = u_0(x) \geq 0, \quad x \in \mathbf{R}, \quad t = 0.$$

Suppose also that $u_0(x)$ of x has compact support (i.e. u_0 is nonzero only inside a closed and bounded subset of $\mathbf{R}$). The solution is given by the convolution

$$u = \int_{\mathbf{R}} u_0(y) \frac{e^{\frac{-|x-y|^2}{4t}}}{\sqrt{4t\pi}} dy, \quad t > 0.$$

Thus if $u_0 \not\equiv 0$, then $u(x,t) > 0$ for all $t > 0$ and $x \in \mathbf{R}$. Hence the support of u is equal to the whole of $\mathbf{R}$ for $t > 0$. Consequently, if u denotes a density function of some distribution of particles say, then a small number of particles must move over arbitrarily large distances in any time interval $[0, \delta t]$.

A further property of Fickian diffusion is illustrated by taking moments of the distribution u. The average particle position is given by

$$\overline{x}(t) = \frac{\int_{\mathbf{R}} xu(x,t)\,dx}{\int_{\mathbf{R}} u(x,t)\,dx}$$

whilst the variance of the distribution is given by the second moment

$$\sigma^2(t) = \frac{\int_{\mathbf{R}} (x - \overline{x})^2 u(x,t)\,dx}{\int_{\mathbf{R}} u(x,t)\,x}.$$

Now $\int_{\mathbf{R}} u(x,t)\,dx$ is constant for solutions of the diffusion equation on $\mathbf{R}$ so, without loss of generality, we take it to be unity.

Differentiating $\overline{x}$ and σ^2 with respect to t, we obtain

$$\begin{aligned} \frac{d\overline{x}}{dt} &= \int xu_t(x,t)dx \\ &= \int xu_{xx}(x,t)dx \\ &= -\int u_x(x,t)dx \\ &= 0 \end{aligned}$$

and

$$\begin{aligned} \frac{d\sigma^2}{dt} &= \int x^2 u(x,t)dx \\ &= -2\int xu(x,t) \\ &= 2\int u(x,t)dx \\ &= 2 \end{aligned}$$

Thus $\overline{x}(t) \equiv \overline{x}(0)$ and $\sigma^2(t) = 2t + \sigma^2(0)$.

The linear growth of the variance is a property that is characteristic of Fickian diffusive processes. However, experiments designed to measure $\sigma^2(t)$ do not always show such behaviour over large time scales, [47].

The alternative to Fickian diffusion is nonlinear, density-dependent diffusion. For example, in section 5.3, we mentioned that populations disperse so as to avoid crowding. The transport part of (5.3.2) is an attempt to model such phenomena via the use of a nonlinear diffusion term.

The use of such nonlinear diffusion in population biology is becoming popular (see [1],[27],[31],[33],[57]), but there are many other areas where it is appropriate. We begin by discussing a few of these.

Consider a gas flowing through a porous medium. Let $\rho(x,t)$ denote the density of the gas. Then conservation of mass implies

$$\theta \rho_t = -\nabla.(\rho \mathbf{v})$$

where θ is the volume fraction filled with gas (called the **porosity** of the medium), and $\mathbf{v}(x,t)$ is the so-called Darcy velocity of the gas. This is simply the velocity the gas would have were it flowing through unoccupied space. The average gas velocity is $\mathbf{v}/\theta$.

Next we have two empirical laws

$$\begin{aligned} \mathbf{v} &= -\frac{k}{\mu}\nabla p, \\ \rho &= \rho_0 p^\gamma. \end{aligned}$$

The first is **Darcy's law** which gives the Darcy velocity, $p(x,t)$ denotes the pressure of the gas, μ is the viscosity of the gas (constant), and k denotes the permeability of the medium (for a given porous material, one would determine k from experimental measurements of $\mathbf{v}$ and p).

The second relation (ρ_0 and γ are positive constants) is called the **equation of state** and relates pressure to density. Eliminating p and $\mathbf{v}$, we obtain

$$\begin{aligned} \rho_t &= \frac{k}{\rho_0^{1/\gamma}\mu\theta}\nabla.(\rho\nabla\rho^{1/\gamma}) \\ &= \frac{k}{\rho_0^{1/\gamma}\mu\theta\gamma m}\Delta(\rho^m) \end{aligned}$$

where $m = 1 + \gamma^{-1}$.

Rescaling time, we obtain

$$\rho_t = \Delta(\rho^m). \qquad (5.4.1)$$

In this setting, (5.4.1) is called the **porous medium equation**.

Next we consider the seepage of moisture through dry soil [74]. The moisture concentration equation is given by

$$c_t + \nabla.\mathbf{J} = 0,$$

and we assume that the flux $\mathbf{J}$ is given by

$$\mathbf{J} = -\kappa(c)\nabla p$$

where the pressure p is due to the capillary potential $h(c)$. We obtain

$$c_t = \nabla.(\kappa(c)\nabla h(c)),$$

a nonlinear diffusion equation (both κ and h are taken to be increasing functions of c, usually given by power laws).

As a further example, consider the spreading of an oil drop on a flat plate. The surface of the oil is given by a function $z = h(x, y, t)$, nonzero over the region covered by the droplet.

The oil is viscous, and we shall assume that the velocity field $\mathbf{u} = (u, v, w)T$ satisfies the **slow flow** approximation to the Navier-Stokes equations (valid in the limit of a small Reynolds number; see Box K). We have

$$\begin{aligned}
p_x &= \mu\Delta u \\
p_y &= \mu\Delta v \\
g + p_z &= \mu\Delta w \\
u_x + v_y + w_z &= 0.
\end{aligned}$$

Here, μ is the viscosity and g the constant of gravity. At $z = 0$ on the flat plate, we have the stationary fluid

$$u = v = w = 0.$$

On the surface $z = h(x, y, t)$, we must have

$$w = h_x u + h_y v + h_t$$

(since a fluid element upon the surface must remain there) and we also assume that there are no shear or normal stresses so that we set

$$\nabla\mathbf{u} + \nabla\mathbf{u}^T = 0,$$

where $z = h(x, y, t)$. We shall also assume that the gas pressure (and hence the surface pressure) is uniform, and, without loss of generality, equal to zero.

Now we assume that h is small, $O(\varepsilon)$ say, relative to the (x, y)-scale. So, we rescale

$$z = \varepsilon \tilde{z}$$
$$h = \varepsilon \tilde{h}.$$

Substituting into all of the above equations, we see that we must rescale

$$u = \varepsilon \tilde{u}$$
$$v = \varepsilon \tilde{v}$$
$$w = \varepsilon^2 \tilde{w}$$
$$p = \varepsilon \tilde{p}$$
$$t = \varepsilon \tilde{t}$$

in order that each equation contains at least two terms balanced in powers of ε.

Using the new variables and dropping the tildes, we retain only the leading terms in all expressions. We have

$$p_x = \mu u_{zz} \tag{5.4.2}$$
$$p_y = \mu v_{zz} \tag{5.4.3}$$
$$g + p_z = 0 \tag{5.4.4}$$
$$u_x + v_y + w_z = 0 \tag{5.4.5}$$

together with

$$u = v = w = 0$$

at $z = 0$ and $p = 0$

$$h_t + h_x u + h_y v = w, \tag{5.4.6}$$

$$\nabla \mathbf{u} + \nabla \mathbf{u}^T = \begin{bmatrix} 0 & 0 & u_z/2 \\ 0 & 0 & v_z/2 \\ u_z/2 & v_z/2 & 0 \end{bmatrix} + O(\varepsilon) = 0$$

at $z = h(x, y, t)$.

Integrating (5.4.4) for p, we have

$$p = g(h - z)$$

so that (5.4.2) and (5.4.3) imply

$$u = \frac{g}{\mu 2} h_x (z^2 - 2h_z)$$
$$v = \frac{g}{\mu 2} h_y (z^2 - 2h_z)$$

where we have used $u = v = 0$ at $z = 0$ and $u_z = v_z = 0$ at $z = h$.

Hence (5.4.5) gives

$$w_z = -\frac{g}{\mu 2}(h_{xx} + h_{yy})(z^2 - 2hz) + \frac{gz}{\mu}(h_x^2 + h_y^2)$$

so

$$w = -\frac{g}{\mu 2}(h_{xx} + h_{yy})(\frac{z^2}{3} - hz^2) + \frac{gz^2}{\mu 2}(h_x^2 + h_y^2).$$

Finally, (5.4.6) implies

$$\begin{aligned} h_t &= \frac{g}{\mu 3}\left(h^3(h_{xx} + h_{yy} + 3h^2(h_x + h_y)^2\right) \\ &= \frac{g}{\mu 3}\left(\frac{\partial}{\partial x}(h^3 h_x) + \frac{\partial}{\partial y}(h^3 h_y)\right) \\ &= \frac{g}{\mu 12}\left(\frac{\partial^2}{\partial x^2} + \frac{\partial^2}{\partial y^2}\right)(h^4) \end{aligned}$$

which is in the form of the porous medium equation (5.4.1) in $\mathbf{R}^2$ with $m = 4$.

Besides the above applications, (5.4.1) has also been applied to the problem of interstellar dispersal of galactic civilizations!

Of course, the term nonlinear diffusion includes any equation of the form

$$u_t = \Delta(\phi(u))$$

for any function ϕ, but the theory is usually restricted to $\phi(u) = u^m$ $(m > 1)$, since the solutions to such equations exhibit the qualitative properties desired. Moreover, there exist group invariant (similarity) solutions for (5.4.1) which are derived as follows, using the method and notation introduced in Box E.

Let $\mathbf{x} \in \mathbf{R}^n$ and $r = \|\mathbf{x}\|$. We seek radially symmetric solutions of

$$u_t = \Delta(u^m).$$

This equation is invariant under the group of transformations

$$\begin{aligned} t &\to \varepsilon t \\ u &\to \varepsilon^q u \\ r &\to \varepsilon^s r \end{aligned}$$

where q, s, and m satisfy

$$q - 1 = qm - 2s.$$

We choose invariants

$$w = \frac{u}{t^q} \qquad \text{and} \qquad y = \frac{r}{t^s}$$

so that

$$u_t = t^{q-1}(qw - sw_y y)$$

and

$$\frac{(r^{n-1}(u^m)_r)_r}{r^{n-1}} = t^{mq-2s}\frac{(y^{n-1}(w^m)_y)_y}{y^{n-1}}.$$

Thus $w(y)$ must satisfy

$$(y^{n-1}(w^m)_y)_y = qy^{n-1}w - sy^n w_y.$$

Choosing $q = -ns$, we may integrate once so that

$$w_y^m = -syw + \text{constant}.$$

This again is integrable when the constant of integration is set to zero. We obtain

$$\frac{m}{m-1}w^{m-1} = \text{constant} - \frac{s}{2}y^2. \tag{5.4.7}$$

We have fixed $q = -ns$ and

$$s = \frac{1}{((m-1)n+2)}.$$

Rewriting the solution in (5.4.7) in terms of u, r, and t, we obtain

$$u = \frac{1}{t^{ns}}\left[\text{constant} - \frac{(m-1)s}{2m}\frac{r^2}{t^{2s}}\right]^{1/(m-1)}. \tag{5.4.8}$$

Our similarity solution is given by the positive part of (5.4.8), that is

$$u = \max\left\{\frac{1}{t^{ns}}\left[\text{constant} - \frac{(m-1)s}{2m}\frac{r^2}{t^{2s}}\right]^{1/(m-1)}, 0\right\}.$$

These one-parameter families of solutions are well-known and provide an illustration of a general property of solutions: the support of solutions to (5.4.1) varies continuously with time. For example, in Figure 5.9, we show the solution $u(x, t)$ for $x \in \mathbf{R}$ at different times $t_1 < t_2 < t_3$.

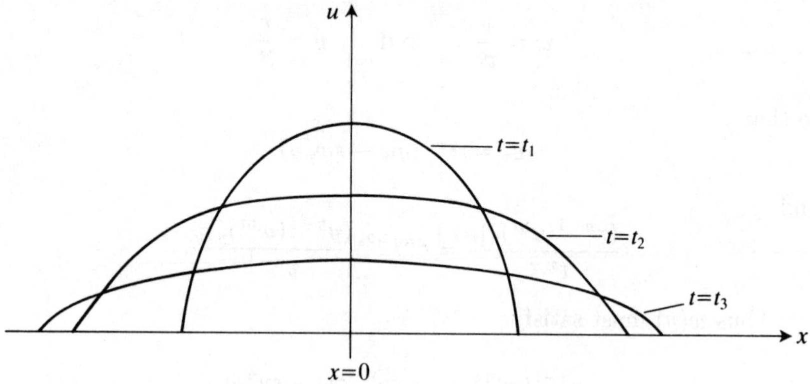

Figure 5.9: A similarity solution for (5.4.1).

Strictly speaking, our solutions are **weak** ones since the derivatives required by the equation may not exist on the boundary of the support of u (see the next section for a further discussion of weak solutions). The questions of existence and uniqueness of solutions are discussed in [64]. For the moment we simply note that each of our similarity solutions, u, is a classical solution wherever it is positive, and the flux (∇u^m) is absolutely continuous everywhere. These properties are generally held by solutions and we refer to [2],[64] and the references therein for a further discussion.

When a nonlinear diffusion term is coupled with a reaction term, we expect to see similar behaviour to that exhibited by standard reaction-diffusion systems. For example, consider the boundary-value problem

$$\begin{aligned} u_t =& (u^m)_{xx} + f(u), \quad x \in (-L, L), \\ & u(\pm L, t) = 0, \quad t \geq 0. \end{aligned}$$

Here, $f(u) = u(1-u)(u-a)$ for some constant $a \in (0,1)$. In order to analyse the long term behaviour of **nonnegative solutions**, u, we seek solutions of the steady-state equation

$$(u^m)_{xx} + f(u) = 0, \quad x \in (-L, L), \quad u(\pm L) = 0. \tag{5.4.9}$$

Multiplying by $(u^m)_x$ and integrating, we obtain

$$\frac{1}{2}(u^m)_x^2 + F(u) = \text{constant}$$

where

$$F(u) = \int_0^u y^{m-1} f(y)\, dy.$$

One may check that if $a > (m+1)/(m+3)$, then $F < 0$ everywhere. Moreover, if $a \in (0, (m+1)/(m+3))$, then $u(x)$ is a nonnegative solution of (5.4.8) if and only if there is some constant $\beta \in (0,1)$, such that $F(\beta) > 0$,

$$|x| = \sqrt{\frac{m}{2}} \int_{u(x)}^{\beta} \frac{y^{m-1} dy}{\sqrt{F(\beta) - F(y)}}$$

for all $x \in (-L, L)$. This is similar to the time-map analysis employed in section 1.4. Notice that $u(0) = \beta$. In particular, β and L must be related via

$$L = \sqrt{\frac{m}{2}} \int_0^{\beta} \frac{y^{m-1} dy}{\sqrt{F(\beta) - F(y)}}.$$

For $0 < (m+1)/(m+3)$, we vary β and sketch the bifurcation curve in the $(u(0), L)$-plane; see Figure 5.10.

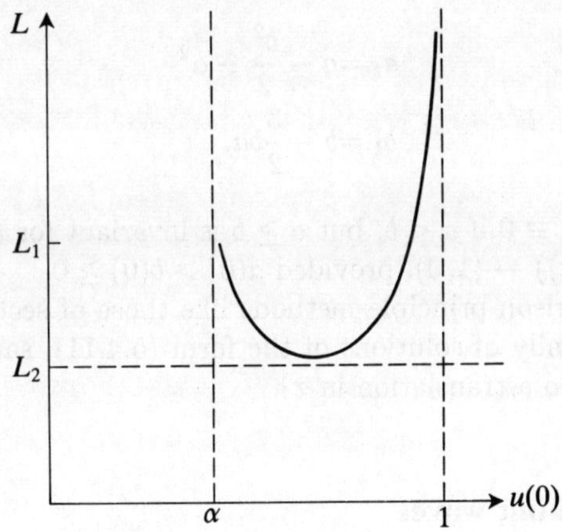

Figure 5.10: Bifurcation diagram for (5.4.9).

For $m > 1$ and $L \in (L_0, L_1)$, there exist two solutions. The lower one vanishes as L increases through L_1. Why? How does this result compare with the case where $m = 1$? (Note that, at $\beta = u^*$ and $L = L_1$, the corresponding solution satisfies $(u^m(L_1))_x = 0$.)

In fact, there are some explicit solutions to nonlinear reaction diffusion equations. However, these are not necessarily group invariant so their detection is more of an art than a science. For example, consider

$$u_t = (u^2)_{xx} + u(1-u), \qquad x \in \mathbf{R}, \quad t \geq 0. \tag{5.4.10}$$

Let us set

$$u = \left[a(t) - b(t)\cosh \alpha x\right]_+ \tag{5.4.11}$$

where $[.]_+$ denotes the nonnegative part of the expression in the brackets, and zero elsewhere. Here, α is some constant and $a(t)$ and $b(t)$ must be chosen.

Substituting (5.4.11) into (5.4.10), we obtain

$$\begin{aligned} a_t - b_t \cosh \alpha x = &- 2ba\alpha^2 \cosh \alpha x + 4\alpha^2 b^2 \cosh^2 \alpha x \\ &- 2\alpha^2 b^2 + a - b \cosh \alpha x \\ &- a^2 + 2ba \cosh \alpha x - b^2 \cosh^2 \alpha x \end{aligned}$$

where $a \geq b \cosh \alpha z$.

Hence we set $\alpha = 1/2$ so that (5.4.11) is a solution of (5.4.10) provided

$$\begin{aligned} a_t =& a - \frac{b^2}{2} - a^2 \\ b_t =& b - \frac{3}{2} ba. \end{aligned} \tag{5.4.12}$$

Notice that $u \equiv 0$ if $a < b$, but $a \geq b$ is invariant for (5.4.12). In fact, as $t > 0$, $(a(t), b(t)) \to (1, 0)$, provided $a(0) > b(0) \geq 0$.

Using comparison principle methods like those of section 1.8, one can show that the family of solutions of the form (5.4.11), satisfying (5.4.12), are stable (modulo a translation in x).

5.5 More travelling waves

The inclusion of nonlinear diffusion terms does little to alter the problems faced in constructing travelling waves, with the exception that we must extend our definition and understanding of plane waves to include waves which have semi-infinite support, say, rather than being classical solutions of the reaction-diffusion equation.

For example, consider

$$u_t = (u^2)_{xx} + u(1-u), \qquad t \geq 0, \quad x \in \mathbf{R}, \tag{5.5.1}$$

where

$$u \to 1 \quad \text{as} \quad x \to -\infty$$
$$u \to 0 \quad \text{as} \quad x \to +\infty.$$

Introducing the travelling variable $z = x - ct$, and writing $u = \phi(z)$, we obtain

$$(\phi^2)_{zz} + c\phi_z + \phi(1-\phi) = 0, \qquad z \in \mathbf{R}. \tag{5.5.2}$$

This has a solution

$$\phi = \left[1 - \exp(z/2)\right]_+, \; c = 1/2 \tag{5.5.3}$$

where $[v(z)]_+ = \max\{v(z), 0\}$.

Notice that (5.5.3) is twice differentiable for $z < 0$, and is a classical solution (i.e. for $x \neq ct$). At $z = 0$, $\phi\phi_z$ is bounded and locally integrable.

Let $w(x)$ be **any** smooth function with compact support in x, for all t in $[0, T]$. Such a solution is called a test function.

Consider the quantity

$$\begin{aligned} P = \int_0^T \int_{\mathbf{R}} -w_x(uu_x) + w_t u + wu(1-u)\,dxdt \\ + \int_{\mathbf{R}} w(x,0)u(x,0) - w(x,t)u(x,t)\,dx. \end{aligned} \tag{5.5.4}$$

It is easy to show that $P \equiv 0$ for all test functions, w, if u is a classical solution of (5.5.1) (integrate by parts).

However, $P \equiv 0$ for other functions u. In particular, $P \equiv 0$ for all suitable test functions w, whenever u is a classical solution, everywhere in $\mathbf{R} \times [0, T]$, except for a finite number of curves across which both u and uu_x are bounded and continuous. We used such a condition to define **weak solutions** of hyperbolic equations in Chapter 3. The same idea holds here and we define u to be a weak solution of (5.5.1) whenever P in (5.5.4) is zero for all test functions, w.

The solutions of the porous media equation discussed in the previous section were weak solutions in the above sense.

It is easy to show that our travelling wave

$$u(x,t) = \phi(x - ct),$$

where ϕ is given by (5.5.3), is a weak solution of (5.5.1).

The method of transition layers used in Chapter 3 is often employed for systems of reaction-diffusion equations having one or more nonlinear diffusion term and an appropriate small parameter. We shall not consider such analysis here and instead concentrate upon a simple method which may be used to reduce the plane wave problem for scalar problems to more standard semi-linear equations [15],[34].

Consider the problem

$$u_t = (D(u)u_x)_x + f(u) \quad \text{for } x \in \mathbf{R},\ t \geq 0. \tag{5.5.5}$$

Suppose this possesses a monotone decreasing travelling wave solution $u = \phi(z)$ where ϕ satisfies

$$\begin{aligned} (D(\phi)\phi_z)_z + c\phi_z + f(\phi) &= 0, \qquad z \in \mathbf{R}, \\ \phi \to 1, \quad & z \to -\infty, \\ \phi \to 0, \quad & z \to +\infty. \end{aligned} \tag{5.5.6}$$

Here, D is continuously differentiable, strictly monotone, and positive on $(0, \infty)$, although we allow $D(0) = 0$; f is Lipschitz continuous on [0,1] and $f(0) = f(1) = 0$.

Without loss of generality, let us suppose ϕ_{zz} exists, $\phi_z < 0$, and $1 \geq \phi > 0$ for $z \in (-\infty, z_0)$ and $\phi \equiv 0$ for $z > z_0$ for some $z_0 \geq 0$ (possibly infinite). Introduce the new independent variable ξ where

$$\frac{d\xi}{dz} = \frac{1}{D(\phi)}, \qquad \xi = 0 \text{ when } z = 0.$$

Now set $\psi(\xi) = \phi(z)$ so that

$$\psi_\xi = \phi_z D(\phi).$$

Multiplying (5.5.6) by $D(\phi)$, we obtain

$$\begin{aligned} \psi_{\xi\xi} + c\psi_\xi + D(\psi)f(\psi) &= 0, \qquad \xi \in \mathbf{R}, \\ \psi \to 1, \quad & \xi \to -\infty, \\ \psi \to 0, \quad & \xi \to \infty. \end{aligned} \tag{5.5.7}$$

Notice that, if $D(0) = 0$, then $\xi \to \infty$ as $z \to z_0$, whence ϕ, and ψ, $\to 0$.

(5.5.7) is of the form of a semilinear travelling wave problem: all of the nonlinearity has passed into the reaction term $D(\psi)f(\psi)$.

One may also proceed in the reverse direction. If $\psi(\xi)$ is a monotone decreasing C^2 solution of (5.5.7), then define z via

$$z = z(\xi) = \int_0^\xi D(\psi(\tilde{\xi}))\, d\tilde{\xi}$$

and set $\phi(z) = \psi(\xi)$.

If $z(\xi) \to z_0$ as $\xi \to \infty$, then ϕ is a classical solution of (5.5.6) for $z < z_0$, and $\phi(z_0) = 0$. Furthermore, if ϕ is extended to be zero for all $z \geq z_0$, then the result is a weak travelling front solution for (5.5.6).

If $D > 0$ for all $\phi \geq 0$, then the above method still holds but with the stronger result that $z(\xi) \to \infty$ as $\xi \to \infty$, so that ϕ is a classical travelling front solution for (5.5.6) with support on the whole of $\mathbf{R}$.

This method is useful since there may be standard results available for (5.5.7) which yield corresponding weak solutions for (5.5.6). For example, if $D = u$ and $f = u(1-u)$, the equation (5.5.7) possesses an explicit solution (see section 1.4) which in turn yields the solution (5.5.3) of (5.5.1).

Exercise 5.1

Consider the radially symmetric similarity solution

$$u = \max\left\{\frac{1}{t^{ns}}\left[k - \frac{s(m-1)}{2m}\frac{r^2}{t^{2s}}\right]^{\frac{1}{(m-1)}}, 0\right\}$$

of the equation

$$u_t = \Delta(u^m), \qquad \mathbf{x} \in \mathbf{R}^n,\ t > 0,$$

which we obtained in section 5.4. Here $r = |\mathbf{x}|$, $k > 0$ is any constant, and

$$s = \frac{1}{((m-1)n+2)}.$$

Show that the variance of the distribution, defined by

$$\sigma^2 = \int_0^\omega r^2 u(r,t) r^{n-1} dr.\omega_n,$$

where ω_n is the area of the $(n-1)$ dimensional unit sphere in n dimensions $(= 2\pi^{\frac{n}{2}}/\Gamma(n/2))$, may be written as

$$\sigma^2 = \frac{2}{t^{2+(m-1)n}} \int_0^{y_0} w(y) y^{(n+1)} dy.\omega_n$$

where $y_0 = \sqrt{k2m/s(m-1)}$. (Hint: the variables y and $w(y)$ are those introduced in section 5.4).

Thus σ^2 grows sublinearly – hence the term *slow diffusion* which is often when this nonlinear equation is used in preference to the linear Fickian diffusion process.

6 Flux driven waves in geochemistry

6.1 Introduction

So far we have been most concerned with waves arising within excitable systems. Such waves propagate via an active threshold mechanism taking place at the fronts or transition layers themselves. Energy is used up in actively exciting successive domains – which is resupplied during a refractory period, following the wave.

There are, though, a large class of reaction-diffusion problems in which waves can only be supported, and propagated, if constituents within the state are brought directly to the reaction zone. In the absence of such a process of advection, all waves would be diffuse, becoming more and more spread out, and patterns could not exist for long times.

In this chapter we shall consider a number of problems concerning flux driven waves. They are typically reaction-diffusion-advection systems. Such situations occur in models of chemical processes and combustion problems for gas and fluid fuels (where there is a considerable literature). The reason for our interest here lies in the fact that such systems typically have some fast (almost instantaneous) processes, and perhaps some kinetic (slower) processes, relative to the transport timescale. Thus, in one spatial dimension, internal transition layer analysis is appropriate in analyzing the wave profiles. However, we shall also employ the ideas of Chapter 4 in order to consider how such waves can behave in two or three dimensions, and their response to local heterogeneity in the advective flow field which supports them.

As a canonical problem we shall consider systems arising in geochemistry. This is a field in which there is considerable need for mathematical input to the assessment of environmental problems. It is rapidly growing, fuelled not only by the development of suitable mathematical tools, but also by the recognition from geochemists themselves that their (traditional) imposition of thermodynamic equilibrium and also of one dimensional symmetry are inadequate. Indeed a new generation of geochemical models is required, and hence so are qualitative techniques for verifying the behaviour of solutions.

We begin by setting out some background, and defining problems in the form of coupled and constrained reaction-diffusion-advection equations. Then we shall discuss the general nature of one dimensional problems when the incoming solute is in disequilibrium with the (virgin) host mineralogy and a sequence of wave fronts will propagate – each with its own characteristic wave speed. Finally we shall consider two and three dimensional problems and modify the geometric theory of Chapter 4 to describe non-planar wave propagation.

6.2 Geochemical systems

In this section we give a rapid introduction to geochemical systems, which are usually relatively large compared to models from physiology or biology, containing many chemical species. However constraints, such as "fast" reactions, can be used to obtain generalized reaction-diffusion systems with many fewer degrees of freedom.

We model a geological environment to be a saturated porous medium, with porosity $\theta(\mathbf{x}, t)$, defined over a suitable domain, Ω. We characterize the reactive chemical species as mobile solutes with concentrations $\mathbf{u}(\mathbf{x}, t) = (u_1, \ldots, u_n)^{\mathrm{T}}$ (measured in, say, mols/unit volume of the pore water) and as immobile minerals with densities $\mathbf{w}(\mathbf{x}, t) = (w_1, \ldots, w_s)^{\mathrm{T}}$ (say mols/unit volume of the medium). We identify two types of chemical reactions: fast reactions, which are typically solute-solute speciation reactions, for example

$$\mathrm{H}^+ + \mathrm{OH}^- \rightleftharpoons \mathrm{H_2O},$$

which are to be equilibrated (in this case we have $[\mathrm{H}^+][\mathrm{OH}^-] = 10^{-13.9}$); and slow reactions, for example the dissolution of calcite

$$\mathrm{Ca}^{2+} + \mathrm{CO}_3^{2-} \rightleftharpoons [\mathrm{CaCO_3}]_{\mathrm{solid}},$$

for which kinetic reaction rates are appropriate. For the moment, the fact that the division exists is more important to us than the details of its definition.

We shall assume that all fast reactions are in instantaneous equilibrium, whereas the slow reactions are to be treated kinetically. Nevertheless, it is useful to consider this situation as an asymptotic limit of the true dynamic system.

Suppose that there are p fast reactions which have reaction terms $\mathbf{F}(\mathbf{u}, \mathbf{w}) = (F_1, \ldots, F_p)^{\mathrm{T}}$ and s slow reactions with reaction rates $\mathbf{S}(\mathbf{u}, \mathbf{w}) = (S_1, \ldots, S_s)$.

For example if the jth reaction is of the form

$$\alpha[u_1] + \beta[u_2] \overset{K}{\leftrightarrow} \gamma[u_3] + \delta[u_4] \tag{6.2.1}$$

then

$$F_j = -a_1^{\alpha} a_2^{\beta} K + a_3^{\gamma} a_4^{\delta}, \tag{6.2.2}$$

where $a_i = a_i(u_i)$ is the activity of the species i (some given function). Typically we take $a_i(u_i) \propto u_i$ for most solute species.

Instantaneous equilibrium of fast reactions requires $\mathbf{F} = 0$ identically. This is equivalent to assuming that the reactions have a rate which is infinite, so that $\mathbf{F}$ approaches zero more rapidly than any other process.

If the jth slow reaction is a precipitation/dissolution reaction of the form

$$\alpha u_1 + \beta u_2 \underset{k_{2,j}}{\overset{k_{1,j}}{\rightleftharpoons}} \gamma w_1 \tag{6.2.3}$$

then the corresponding slow kinetic reaction term is

$$S_j = -k_{1,j} a_1^{\alpha} a_2^{\beta} + k_{2,j} A(w_1), \tag{6.2.4}$$

where $k_{1,j}$ and $k_{2,j}$ are the rates of the forward and back reactions and A is the activity of the mineral species (w_1), which depends upon specific surface area (grain size, etc.). Most simply it might be a constant for $w_1 > 0$.

Consider the dynamics of the solute species, assuming that the jth fast reaction has reaction rate R_j (a very large positive constant). Conservation of mass implies that

$$(\theta \mathbf{u})_t = \mathcal{L}(\mathbf{u}) + \theta B_s \mathbf{S}(\mathbf{u}, \mathbf{w}) + \theta B_F R \mathbf{F}(\mathbf{u}, \mathbf{w}). \tag{6.2.5}$$

Here $\mathcal{L}(\mathbf{u})$ denotes the net rate of increase in mass per unit volume due to transport of solutes (diffusion and advection); θ is the porosity of the porous medium; $R = \mathrm{diag}(R_1, \ldots, R_p)$ is a $(p \times p)$ diagonal matrix of fast reaction rates, and B_F and B_S are $(n \times p)$ and $(n \times s)$ matrices containing the stoichiometric coefficients of the fast and slow reactions respectively. For example, assuming F_j is given by (6.2.2), the jth column of B_F must be of the form $(\alpha, \beta, -\gamma, -\delta, 0, \ldots, 0)^{\mathrm{T}}$, since there is a net loss of u_1 to u_4 in proportion to α, β, $-\gamma$ and $-\delta$ respectively as the reaction takes place.

We consider (6.2.5) in the limits $R_1, \ldots, R_p \to \infty$. In doing so we must eliminate the $R\mathbf{F}$ terms and so obtain a new system governing the $n - p$ degrees of freedom available in the limit.

We introduce an $n \times (n - p)$ matrix C such that $C^{\mathrm{T}} B_F = 0$. This is arranged by choosing columns of C to be a linearly independent set of vectors each of which is orthogonal to the p-dimensional space spanned by the columns of B_F. Let $\mathbf{c}_i$ denote the ith column of C. The quantities $U_i = \mathbf{c}_i \cdot \mathbf{u}$, simple linear combinations of the solute species, are of central interest since these are conserved with respect to all of the fast reactions: notice that the $\mathbf{c}_i$s are completely determined by the stoichiometry of the fast reactions. Hence the U_is $(i = 1, \ldots, n-p)$ provide a set of coordinates within which the fast reactions are eliminated and the slow kinetics and transport processes can be considered. Premultiplying (6.2.5) by C^{T}, and setting $\mathbf{U} = (U_1, \ldots, U_{n-p}) = C^{\mathrm{T}} \mathbf{u}$, we obtain

$$(\theta \mathbf{U})_t = C^{\mathrm{T}} \mathcal{L}(\mathbf{u}) + \theta C^{\mathrm{T}} B_S \mathbf{S}(\mathbf{u}, \mathbf{w}). \tag{6.2.6}$$

Thus the term $R\mathbf{F}(\mathbf{u},\mathbf{w})$ has been eliminated and (6.2.6) is valid as any $R_i \to \infty$: that is in the limit where the fast reactions are considered as instantaneous.

Hence at any time $t > 0$, we solve (6.2.6) together with the algebraic system

$$0 = \mathbf{F}(\mathbf{u},\mathbf{w}) \tag{6.2.7}$$

$$\mathbf{U} = C^{\mathrm{T}}\mathbf{u} \tag{6.2.8}$$

which yields $\mathbf{u}$ in terms of $\mathbf{U}$ (and trivially vice-versa).

In general (6.2.6) must be coupled with the equations for the mineral species:

$$\mathbf{w}_t = \tilde{B}_S S(\mathbf{u},\mathbf{w}) \quad (\mathbf{w} > 0), \tag{6.2.9}$$

where $\tilde{B}_S$ denotes the stoichiometry matrix for the mineral precipitation and dissolution reactions.

If the transport processes, $\mathcal{L}(\mathbf{u})$, include advection then these equations are augmented by an equation for fluid flow, say

$$\theta_t = -\nabla \cdot [\mathbf{v}] \tag{6.2.10}$$

(conservation of mass of water), where $\mathbf{v}$ is the Darcy velocity. If $\theta_t \neq 0$, that is, there are porosity changes due to the altering mineralogy, then the system is said to be fully coupled and (6.2.10) is augmented by an equation of the form $\theta = g(\mathbf{w},\mathbf{u},\ldots)$. This defines the porosity in terms of the mineralogy. Hence as reactions change the mineralogy, fluid is displaced which leads to further transport of reactants and further reaction: hence the term **fully coupled**.

In the following sections we consider a number of applications of this class of equations.

6.3 Single wave fronts

We consider the reactive transport of a single aqueous species, with concentration u, through a saturated porous media, generalizing Exercise 3.2 by fully coupling the mineralogy to the flow and allowing for a kinetic reaction. To be more precise, we consider

$$\begin{aligned}
(\theta u)_t &= \nabla \cdot [D\nabla u - u\mathbf{v}] - \frac{F(u,q)}{\delta},\\
q_t &= \frac{R(u,q)}{\delta},\\
\theta_t &= -\nabla \cdot [\mathbf{v}],\\
\theta &= g(u,q,\ldots).
\end{aligned} \tag{6.3.1}$$

Here, as before, θ is the porosity, D is the dispersion/effective dispersion coefficient, $\mathbf{v}$ is the Darcy velocity, and q represents the density of the solid or adsorbed phase of the species, assumed immobile. For a **reversible reaction** we may assume there exists a continuously differentiable function f such that $F(u, f(u)) = 0$, for all $u \geq 0$, so that $q = f(u)$ represents that chemical equilibrium. For example we may have a Langmuir isotherm:

$$q = f(u) = \frac{u}{a + bu}$$

for suitable constants a and b (see Exercise 3.2), so that R might have the form

$$F(u, q) = h(u, q)(f(u) - q).$$

If the reaction is **irreversible**,

$$c + \text{host rock} \rightarrow q + \text{other products},$$

then the analysis given at the end of section 3.4 applies. For both the reversible and the irreversible assumptions, the conservation of mass across any plane wave front implies that it moves with a characteristic velocity equal to the average fluid velocity retarded by a factor taking into account the loss of solute species, u, at the reaction zone: the details of the transition profile depend on the reaction terms though. These ideas will be extended to more general systems in section 6.4. We shall analyze a simple irreversible reaction further when we develop a geometric theory for non-planar flux waves in section 6.5.

We first consider (6.3.1) in the limit $\delta \rightarrow 0$; that is, assuming that the chemistry equilibrates on a faster timescale than the transport processes. We obtain a single nonlinear equation

$$(\theta u + f(u))_t = \nabla \cdot [D \nabla u - u\mathbf{v}],$$
$$\theta_t + \nabla \cdot \mathbf{v} = 0.$$

We allow the porosity θ to depend upon the solute and hence the mineral chemistry, and assume this change may be parameterized by u alone (since $w = f(u)$) to write

$$\theta = g(u)$$

for some continuously differentiable function $1 > g > 0$ for all $u \geq 0$.

Consider the existence of one dimensional (planar) travelling waves. Introducing a travelling variable $z = x - \omega t$, for some wave speed ω, and writing $u = U(z), \mathbf{v} = V(z)\mathbf{i}$, we have

$$0 = (g(U)DU_z - UV)_z + \omega(g(U)U + f(U))_z$$
$$0 = \omega g(U)_z - V_z.$$

As $z \to -\infty$ we set $U = u_{-\infty}, V = v_{-\infty}$, the inlet concentration and inlet Darcy velocity at the back of the wave. As $z \to +\infty$ we assume $U = 0$. Solving, we obtain

$$\omega = \frac{u_{-\infty}v_{-\infty}}{(g(u_{-\infty})u_{-\infty} + f(u_{-\infty}))},$$
$$V(z) = \omega(g(U(z)) - g(u_{-\infty})) + v_{-\infty},$$
$$\frac{v_{-\infty}z}{D} = (g(u_{-\infty})u_{-\infty} + f(u_{-\infty})) \int_{\frac{u_{-\infty}}{2}}^{U(z)} \frac{g(u)}{(uf(u_{-\infty}) - u_{-\infty}f(u))} du.$$

Hence a travelling wave solution exists if and only if $f(u) > f(u_{-\infty})u/u_{-\infty}$ for all $u \in (0, u_{-\infty})$, a generalization of the result obtained in Exercise 3.2.

Next we shall relax our assumption that $\delta \to 0$ in (6.3.1). We assume that the reactive formation of the immobile species, q, results in an increase in porosity of the media due to the dissolution of some mineral reactant in the host rock; and that the stage of the reaction can be stated in terms of the local porosity, θ, which changes from $\theta_{+\infty}$, the initial porosity, to $\theta_{-\infty}$, the inlet porosity, where $\theta_{-\infty} > \theta_{+\infty}$. Thus we solve for (u, θ) rather than (u, q). The change in porosity is also accompanied by an associated change in permeability so that the reacted medium is more permeable. In a later section we shall show that the plain waves we are considering may lose stability, resulting in fingering patterns, for systems such as these, whether or not the reaction is reversible.

We let $P(\mathbf{x}, t)$ denote the fluid pressure and apply Darcy's law for the flow as a function of the altering permeability distribution.

We consider the model

$$\{\theta u + \theta\rho\}_t = \nabla \cdot \{D\nabla u + u\kappa\nabla P\},$$
$$\theta_t = k(\theta_{-\infty} - \theta)^{2/3}u,$$
$$\theta_t = -\nabla \cdot \{\mathbf{v}\}, \ \ \mathbf{v} = -\kappa\nabla P.$$

Here $\kappa = \kappa(\theta)$ denotes the permeability of the medium; the constant ρ is the initial molar density of the unreacted mineral (assumed uniform); k is a rate constant for the reaction. The term $(\theta_{-\infty} - \theta)^{2/3}$ is proportional to the density of the unreacted mineral species; and D is the dispersion / effective dispersion coefficient of the solute.

We shall assume that the solute is supplied with the incoming fluid at $x = -\infty$, and that flow is initially in the x-direction. Ahead of the reaction, we have $u = 0$ and $\theta = \theta_{+\infty}$, the initial mineral distribution being uniform and unreacted. As $x \to -\infty$, behind the reaction zone, we shall assume that $u = u_{-\infty}$ (the inlet solute concentration), $\theta = \theta_{-\infty}$ (the inlet porosity), and that the Darcy velocity is given by $\mathbf{v} = (v_{-\infty}, 0)^{\mathrm{T}}$, where $v_{-\infty} = \kappa P_x$, acting uniformly in the x-direction.

Again introducing a travelling coordinate $z = x - \omega t$, for some wave speed ω, we analyze the system in a moving frame seeking a (unique) planar travelling wave front moving the state of the reaction from $(u, \theta) = (0, \theta_\infty)$ up to $(c_{-\infty}, \theta_{-\infty})$.

Writing $u = U(z),\ P = P(z),\ \theta = \Theta(z)$, we obtain the system

$$\left[DU_z + \kappa P_z U + \omega\Theta(U + \rho)\right]_z = 0 \tag{6.3.2}$$

$$\omega\Theta_z = -k(\theta_{-\infty} - \Theta)^{2/3}U \tag{6.3.3}$$

$$0 = \left[\kappa P_z + \omega\Theta\right]_z. \tag{6.3.4}$$

Subject to the boundary conditions

$$\left.\begin{aligned} -\kappa P_z &= v_{-\infty} \\ \Theta &= \theta_{-\infty} \\ U &= u_{-\infty} \end{aligned}\right\} \quad \text{as } z \to -\infty, \tag{6.3.5}$$

$$\left.\begin{aligned} \Theta &= \theta_{+\infty}(< \theta_{-\infty}) \\ U &= 0 \end{aligned}\right\} \quad \text{as } z \to +\infty. \tag{6.3.6}$$

Applying the conditions as $z \to -\infty$ to (6.3.4), we obtain

$$\kappa P_z = -\omega\Theta + \theta_{-\infty}\omega - v_{-\infty}. \tag{6.3.7}$$

Similarly applying the conditions at $z \to -\infty$ to (6.3.2) and (6.3.3), we obtain two first order equations:

$$DU_z = U(v_{-\infty} - \theta_{-\infty}\omega) - \omega\rho(\Theta - \theta_{+\infty}), \tag{6.3.8}$$

$$\omega\Theta_z = -k(\theta_{-\infty} - \Theta)^{2/3}U, \tag{6.3.9}$$

where we have substituted for P_z in (6.3.2) using (6.3.7).

The wave speed ω is fixed now since as $z \to -\infty$, (6.3.8) implies

$$\omega = \frac{v_{-\infty}u_{-\infty}}{u_{-\infty}\theta_{-\infty} + \rho(\theta_{-\infty} - \theta_{+\infty})}.$$

Hence, again, ω is independent of the details of the reaction.

Linearizing (6.3.8) and (6.3.9) about the rest state $(U, \Theta) = (0, \theta_{-\infty})$ the resulting (2×2) matrix system has characteristic polynomial

$$p(\lambda) = \lambda^2 - \lambda\frac{(v_{-\infty} - \theta_{-\infty}\omega)}{D} - (\theta_{-\infty} - \theta_{+\infty})^{\frac{2}{3}}\frac{k\rho}{D}.$$

Since $p(0) < 0$ there is always one negative real eigenvalue, λ_-, say, and one positive real eigenvalue. It is easy to check that λ_- corresponds to a stable manifold in the (U, Θ) plane, with asymptotic slope given by

$$\frac{d\Theta}{dU} = \frac{(\theta_{-\infty} - \theta_{+\infty})^{\frac{2}{3}} k}{\omega |\lambda_-|}.$$

Next consider the flow defined by (6.3.8) and (6.3.9) in the (U, Θ) plane. Consider the triangle

$$T = \left\{ (U, \Theta) : 0 < U < u_{-\infty}, \theta_{+\infty} + U \frac{v_{-\infty} - \theta_{-\infty}\omega}{\rho\omega} < \Theta < \theta_{-\infty} \right\}.$$

We may check that all trajectories inside T must emanate from the rest point $(U, \Theta) = (u_{-\infty}, \theta_{-\infty})$. This follows directly from the fact that no trajectories may enter T from outside at any point on the boundary ∂T, see Figure 6.1.

Now, the stable manifold at $(U, \Theta) = (0, \theta_{+\infty})$ arrives at the rest point from the interior of T since

$$\frac{(\theta_{-\infty} - \theta_{+\infty})^{\frac{2}{3}} k}{\omega |\lambda_-|} > \frac{\theta_{-\infty}(v - \omega)}{\rho\omega},$$

that is, the stable manifold has a steeper slope than the diagonal part of the boundary of T.

This last inequality follows by observing that it must hold for all λ_- such that

$$0 < -\lambda_- < \frac{\rho k (\theta_{-\infty} - \theta_{+\infty})^{\frac{2}{3}}}{(v - \theta_{-\infty}\omega)} \equiv -\lambda^*, \text{ say;}$$

and that $p(\lambda^*) = \lambda^{*2} > 0$. Thus since p is negative for $\lambda \in (-\lambda_-, 0)$, we must have

$$0 < -\lambda_- < -\lambda^*,$$

as required.

Hence the stable manifold at $(U, \Theta) = (0, \theta_{+\infty})$ must emanate from $(U, \Theta) = (u_{-\infty}, \theta_{-\infty})$, and this trajectory corresponds to a unique heteroclinic wave solution satisfying the required conditions at $z = \pm\infty$. Monotonicity follows trivially since neither flow component changes sign in T, again see Figure 6.1.

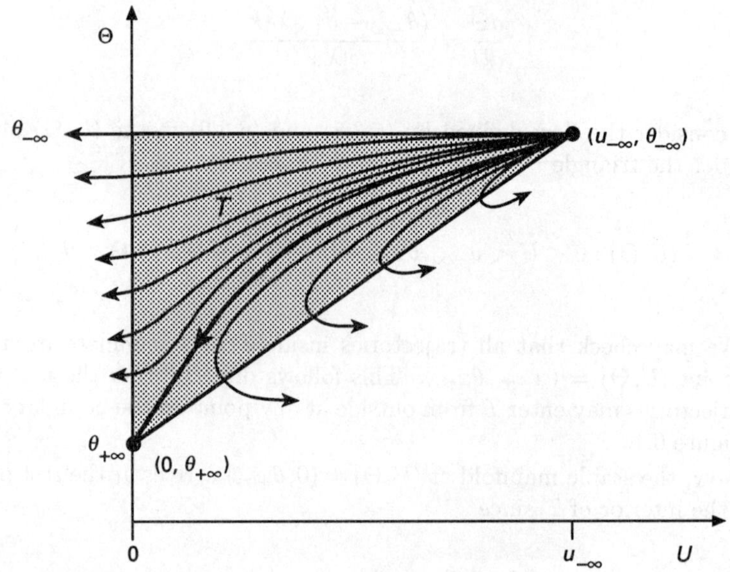

Figure 6.1: Heteroclinic orbit in the (U, Θ) plane.

6.4 Multiple wave fronts

In [32], [56] and [60] some examples of one dimensional flux-driven geochemical waves are discussed. Here we shall consider a generic class of problems in the form of constrained differential equations. We shall show that wave solutions arising due to the advective influx of the solute species in disequilibrium with the host medium are quite generally in the form of a sequence of successive sharp fronts, each moving with its own characteristic speed.

We assume that the dispersion/diffusion coefficients become small, and consequently transitions can be abrupt. Given the sharp fronts observed within natural geochemical systems, see Figure 6.2, this is a natural limit to take. The resulting limiting system is an advection-reaction problem: it is hyperbolic, rather than parabolic, and possesses solutions in the form of so-called shock waves. Much is known about this phenomena, in particular when systems of conservation laws yield multi-shock solutions in order to account for the transition of the state from one ambient (unreacted) equilibrium to another (reacted) equilibrium, even for nonlinear systems.

Figure 6.2: A redox front: oxidizing region shows darker around a water bearing fracture.

Of course, we are only interested in those shock solutions which are obtained as the limit of ever sharpening wave front solutions to the original problem, as the dispersion/diffusion coefficients tend to zero, see section 3.5. The key point is that whether one considers the full *smoothed* equations, or the limiting hyperbolic system, we expect the solutions to break up into a number of distinct sharp transitions, separating zones in chemical equilibrium: very locally each transition will be smooth in reality, though the determination of its evolution is a matter of imposing *conservation of flux* conditions across narrow frontal zones.

We consider a one dimensional porous medium with fluid flow along the x-axis. Initially, the geochemistry is in equilibrium; then we allow an incoming flux of solute species to react with the medium to alter the mineralogy to a new equilibrium.

A classical initial value, or Riemann, problem would specify all species at some time, say $t = 0$. But here we only specify these partially, for example by fixing the incoming solutes only, allowing the reacted mineralogy to be constrained only by the need for equilibrium. Similarly ahead of the reaction, the mineralogy may be fixed but the outflowing solute species unknown. Hence we consider the changes produced by propagating reaction fronts and the overall initial conditions may not necessarily be known *a priori*. This will follow from our development below.

Consider a one dimensional problem of the type introduced in section 6.2. As before we let $\mathbf{u}(x,t) \in \mathbf{R}^n$ denote the distribution of n reactive solute species (mols per unit fluid volume), $\mathbf{q}(x,t) \in \mathbf{R}^s$ denote the distribution of s reactive mineral species (mols per unit *in situ* rock volume); $\theta = g(\mathbf{q})$ be the porosity, a given function of the local mineralogy $\mathbf{q}$; and $v(x,t)$ denote the Darcy fluid velocity. We shall assume there are p fast reactions, all instantaneously equilibrated, on the timescale associated with transport. The stoichiometry of each reaction can be described by an $(n+s) \times p$ matrix of coefficients, B say; the jth column contains the stoichiometric coefficients for the $\mathbf{u}$ and $\mathbf{q}$ species in the jth reaction. Applying conservation of mass we have a system of the form, say,

$$\begin{bmatrix} \theta \mathbf{u}_t \\ \mathbf{q}_t \end{bmatrix} = \begin{bmatrix} \varepsilon \mathbf{u}_{xx} - v\mathbf{u}_x \\ 0 \end{bmatrix} + B\mathbf{F}(\mathbf{u}, \mathbf{q}).$$

Recall that a matrix C is of dimension $(n+s) \times (n+s-p)$ and rank $(n+s-p)$, and is chosen so that the transpose annihilates the stoichiometry matrix, B, associated with the p-equilibrated reactions: $C^{\mathrm{T}} \cdot B = 0$ (the columns of C span the null space of $B \cdot B^{\mathrm{T}}$).

Following section 6.2, we have:

$$C^{\mathrm{T}} \cdot \left\{ \begin{bmatrix} \theta \mathbf{u}_t \\ \mathbf{q}_t \end{bmatrix} + \begin{bmatrix} v\mathbf{u}_x - \varepsilon \mathbf{u}_{xx} \\ 0 \end{bmatrix} \right\} = 0 \qquad x \in \mathbf{R},\ t > 0;$$

conservation of (the incompressible) fluid mass:

$$\theta_t + v_x = 0, \qquad x \in \mathbf{R},\ t > 0;$$
$$\theta = g(\mathbf{q});$$

geochemical equilibrium:

$$\mathbf{F}(\mathbf{u}, \mathbf{q}) = 0, \qquad x \in \mathbf{R},\ t > 0;$$

for the reaction mapping $\mathbf{F} : \mathbf{R}^n \times \mathbf{R}^s \to \mathbf{R}^p$.

Here $\varepsilon > 0$ is the dispersion/diffusion rate of the solute species, which we may assume is small; we shall assume $v > 0$ so that fluid flow takes place in the positive x-direction, and that the geochemical wave is initiated by a discontinuity (or step) in the conditions at $x = 0$, $t = 0$.

We consider the evolution of this system from step initial conditions imposed on the solutes and minerals

$$\mathbf{u} = \mathbf{u}_{+\infty} \qquad x > 0,\ t = 0;$$
$$\mathbf{u}_{-\infty} \qquad x < 0,\ t = 0;$$

$$\mathbf{q} = \begin{cases} \mathbf{q}_{+\infty} & x > 0,\ t = 0; \\ \mathbf{q}_{-\infty} & x < 0,\ t = 0. \end{cases}$$

Here $\mathbf{u}_{-\infty}$ and $\mathbf{q}_{-\infty}$ denote the inlet concentrations: $\mathbf{u}_{+\infty}$ and $\mathbf{q}_{+\infty}$ the unreacted concentrations ahead of the wave. In practice they may not be all known (see below) but must satisfy the equilibrium constraints

$$\mathbf{F}(\mathbf{u}_{\pm\infty}, \mathbf{q}_{\pm\infty}) = 0.$$

For example, the inlet solute concentrations, $\mathbf{u}_{-\infty}$, may be known, though there be no knowledge of the outlet concentrations, $\mathbf{u}_{+\infty}$, ahead of the reaction zones. Similarly the unreacted mineralogy, $\mathbf{q}_{+\infty}$, present ahead of the wave may be known, whereas the composition of the reacted mineralogy, $\mathbf{q}_{-\infty}$, behind the the wave resulting from the reactions is to be determined.

The inlet fluid flux v is given behind the wave

$$v \to v_{-\infty} \quad \text{as } x \to -\infty,$$

whereas its value ahead of the wave depends upon the overall change in porosity, which depends upon the mineralogy.

If $\varepsilon = 0$, and we consider the advection-reaction problem, this is a classical Riemann problem for a system composed of a conservation law [44]. We expect a solution composed of successive shock waves. Following some specific previous examples [32], [14], we anticipate that the solution to the current problem will consist of a smoothed version of such waves, with diffusion/dispersion relevant in a small, $O(\varepsilon)$, transition layer located about each shock.

We therefore seek a solution of the form of a sequence of m sharp fronts, depicted in Figure 6.3, dividing zones where $(\mathbf{u}, \mathbf{q})$ are constant.

We seek intermediate states $(\mathbf{u}_1, \mathbf{q}_1), (\mathbf{u}_2, \mathbf{q}_2), \ldots, (\mathbf{u}_{m-1}, \mathbf{q}_{m-1})$; together with the wave speeds: $\omega_1 > \omega_2 > \cdots > \omega_m$.

However $\mathbf{u}_{\pm\infty}, \mathbf{q}_{\pm\infty}$ are not fully determined, so let r denote the number of degrees of freedom which remain after imposing the conditions

$$\mathbf{F}(\mathbf{u}_{\pm\infty}, \mathbf{q}_{\pm\infty}) = 0.$$

Thus we have a total of $(n+s)(m-1) + 2m + r$ unknowns.

At the ith front, setting $\varepsilon z = x - \omega_i t$, so that z is a local scaled coordinate, and writing $\mathbf{u} = \mathbf{u}(z)$ etc., we obtain

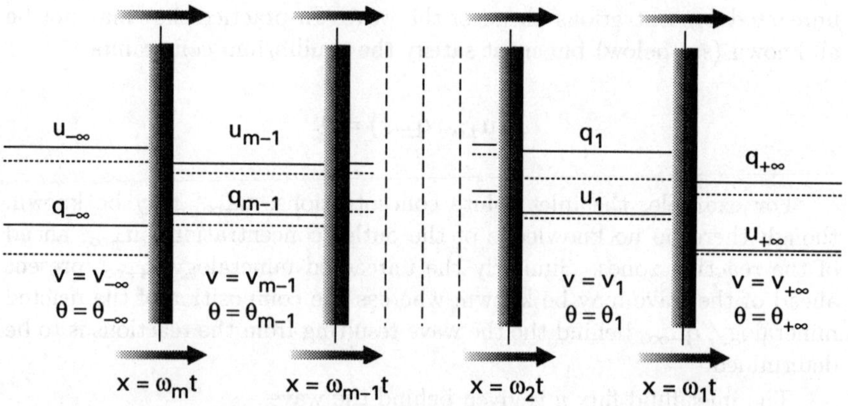

Figure 6.3: Successive fronts in one dimension, dividing equilibrated states, each moving with its own characteristic speed.

$$C^{\mathrm{T}} \cdot \left\{ \begin{bmatrix} v\mathbf{u} - \mathbf{u}_z \\ 0 \end{bmatrix} - \begin{bmatrix} \omega_i \theta \mathbf{u} \\ \omega_i \mathbf{q} \end{bmatrix} \right\}_z = 0.$$

Integrating and imposing the quiescent constant conditions for $z > 0$ and large, and $z < 0$ and large, we obtain

$$C^{\mathrm{T}} \begin{bmatrix} (v_{i-1} - \omega_i \theta_{i-1}) I_n & 0 \\ 0 & -\omega_i I_s \end{bmatrix} \begin{bmatrix} \mathbf{u}_{i-1} \\ \mathbf{q}_{i-1} \end{bmatrix} = C^{\mathrm{T}} \begin{bmatrix} (v_i - \omega_i \theta_i) I_n & 0 \\ 0 & -\omega I_s \end{bmatrix} \begin{bmatrix} \mathbf{u}_i \\ \mathbf{q}_i \end{bmatrix},$$

where I_n and I_s denote the n and s dimensional identities respectively, with the convention that $(\mathbf{u}_0, \mathbf{q}_0) = (\mathbf{u}_\infty, \mathbf{q}_\infty)$ and $(\mathbf{u}_m, \mathbf{q}_m) = (\mathbf{u}_{-\infty}, \mathbf{q}_{-\infty})$. This amounts to total conservation of flux across the transition fronts. We also have

$$v_i - \omega_i \theta_i = \text{constant},$$

where $\theta_i = g(\mathbf{q}_i)$.

Within each inter-front zone we must have local equilibrium so that

$$\mathbf{F}(\mathbf{u}_i, \mathbf{q}_i) = 0 \qquad i = 1, \ldots, m-1,$$

(where $\mathbf{u}_m = \mathbf{u}_\infty$, $\mathbf{q}_m = \mathbf{q}_\infty$ again).

Hence we have a total of $m(n+s-p+1)$ jump conditions, and $p(m-1)$ equilibrium conditions, giving $m(n+s+1)-p$ conditions in all. Equating this with the number of unknown variables, we have

$$m = n + s - p - r. \tag{6.4.1}$$

Since the conservation laws are nonlinear these cannot be a sufficient condition for the existence of shock solutions, so (6.4.1) remains merely necessary: nevertheless it is useful in practice when such system solutions are sought.

The solutions which we have considered above constitute the "outer" solutions which must be matched with the wave profiles inside the fronts. Their precise nature depends upon the form of the reaction kinetics, and indeed, the timescales appropriate to both the dispersal processes and the reactions, which were considered here to be negligible.

Two recent works, [56] and [4], give a number of examples based upon well studied geochemical systems, which give rise to successive fronts propagating at various speeds. In particular the consideration of the oxidation of pyrite in the presence of k-feldspar leads to systems of equations in the form discussed above, with $m = 5$ successive sharp fronts. These examples are used to motivate a consideration of advection-reaction models in one spatial dimension. This approach follows the evolution of the geochemistry along a single streamline or "reaction path", and can incorporate kinetics as well as equilibrated reactions. However it is not easy to generalize such results to two or more dimensions whilst allowing for heterogeneous fluid flow.

6.5 Non-planar flux driven waves

We have seen how geochemical transport problems generically yield solutions in the form of successive wave fronts. In this section we consider how such fronts evolve in heterogeneous two or three dimensional media.

We begin by considering a single solute-solid dissolution reaction in nondimensionalized terms, with $u(\mathbf{r}, t)$ denoting the concentration of an inlet solute phase, and $q(\mathbf{r}, t)$ denoting the density of a reactive mineral in the host rock. The analysis below can be carried out in a very straightforward way for reactions characterized by isotherms (it is somewhat simpler). So here we shall illustrate it with reference to a non-isotherm (irreversible) reaction.

We have

$$\begin{aligned} \theta u_t &= \varepsilon \Delta u - \mathbf{v} \cdot \nabla u - \frac{1}{\delta^2} F \\ q_t &= -\frac{1}{\delta^2} F \end{aligned} \tag{6.5.1}$$

where the reaction term F satisfies

$$F \geq 0; \quad F = 0 \Leftrightarrow u = 0 \text{ or } q = 0$$

In this example we assume constant porosity. The rate constant δ^{-2} is written in this form since we wish to consider the case where $\delta \to 0$, that is, the reaction becomes fast. In fact we have two small constants: ε, the dispersivity, and δ; and as a consequence, our single transition layer is really a diffusive layer (of width $O(\varepsilon)$) and a further inner reactive layer (of width $O(\delta)$), on the assumption $\delta \ll \varepsilon$.

For ε small, and δ even smaller, the medium will be divided into three regions: an unreacted zone, downstream, where $u \approx 0$, $q \approx q_{+\infty}$; a reacted zone, upstream, where $u \approx u_{-\infty}$, the inlet concentration, and $q = 0$; and a fine transition layer dividing the two. In fact, considering one dimensional wave propagation, the system possesses a travelling wave front, propagating with constant speed, for which the transition layer may be shown directly to be of width $O(\varepsilon)$.

We shall construct asymptotic solutions of (6.5.1), valid in the limit $\varepsilon \to 0$, following Chapter 4, exploiting the existence of the sharp transition in order to apply singular perturbation theory and obtain an equation governing the geometric evolution of **non-planar waves** in two and three spatial dimensions.

Suppose that (u, q) is a solution of (6.5.1) and that there is an oriented surface M moving through $\mathbf{R}^3$ such that $u \approx 0$, $q \approx q_{+\infty}$, ahead of it (in the oriented normal direction) and $u \approx u_{-\infty}$, $q \approx 0$, behind it. These are distinct steady states of (6.5.1). At M we must locate a transition layer so as to match these two constant "outer" solutions. In doing so, we must consider the evolution (motion) of M and also determine the transitional profile.

As in Box I, Chapter 4, for each time t, let $\mathbf{r}^*(\eta, \lambda, t)$ range over a subset, or patch, M_0, of M as real parameters η and λ vary over some suitable intervals. We introduce a (scaled) normal coordinate ξ via the mapping

$$\mathbf{r} = \mathbf{r}^*(\eta, \lambda, t) + \varepsilon\xi\mathbf{n}(\mathbf{r}^*) + O(\varepsilon^2) \tag{6.5.2}$$

so that for t fixed the mapping $(\xi, \eta, \lambda) \to \mathbf{r}$ is invertible close to M_0, and set

$$u = \psi(\xi), \qquad q = \chi(\xi) \tag{6.5.3}$$

in a neighbourhood of M_0, and demand that u and q satisfy (6.5.1) together with the conditions

$$\begin{aligned} &\psi \to 0, \ \chi \to q_{+\infty} \quad &&\text{as } \xi \to \infty \\ &\psi \to u_{-\infty}, \ \chi \to 0 \quad &&\text{as } \xi \to -\infty. \end{aligned}$$

As in Box I, we may calculate directly the derivatives in (6.5.1) as follows:

$$u_t = -\psi_\xi \frac{N}{\varepsilon}, \quad q_t = -\chi_\xi \frac{N}{\varepsilon}$$

where $N = \mathbf{r}_t^* \cdot \mathbf{n}$ is the normal velocity of the surface;

$$\Delta u = \frac{\psi_{\xi\xi}}{\varepsilon^2} + \frac{\psi_\xi K}{\varepsilon},$$

where

$$K = -\left(\frac{\mathbf{r}_{\eta\eta}^*}{|\mathbf{r}_\eta^*|^2} + \frac{\mathbf{r}_{\lambda\lambda}^*}{|\mathbf{r}_\lambda^*|^2}\right) \cdot \frac{\mathbf{r}_\xi^*}{\varepsilon}$$

is twice the mean curvature of the surface; and

$$\nabla \mathbf{u} = \psi_\xi \frac{\mathbf{n}(\mathbf{r}^*)}{\varepsilon}.$$

Substituting these into (6.5.1), we have, to lowest order in ε:

$$0 = N\chi_\xi/\varepsilon - F/\delta^2$$

$$0 = \varphi_{\xi\xi} - \mathbf{v} \cdot \mathbf{n}\varphi_\xi + N(\theta\varphi - \chi)_\xi + \varepsilon K \varphi_\xi.$$

This has an "outer" transition layer solution

$$\xi > 0 \begin{cases} \chi = 0 \\ \varphi = u_{-\infty}(1 - e^{(\mathbf{v}\cdot\mathbf{n} - N\theta - \varepsilon K)\xi}) \end{cases}$$

$$\xi > 0 \begin{cases} \chi = q_{+\infty} \\ \varphi = 0 \end{cases}.$$

Note that from the left φ approaches 0 with non-zero (negative) slope as $\xi \to 0$; but in fact φ must be smooth so we must produce an inner expansion, valid near $\xi = 0$, matched to the outer expansion.

We rescale by introducing

$$\zeta = \frac{\xi}{\delta}$$

and

$$\delta U(\zeta) = \varphi, \quad Q(\zeta) = \chi$$

(since both ξ and φ are small near $\xi = 0$). We have, to the lowest order in δ,

$$(U_\zeta - NQ)_\zeta = 0 \Leftrightarrow (U_\xi - NQ) = \text{constant},$$

and another equation for Q_ζ. Ahead of this inner wave we match to the outer transition layer solution (φ, χ):

$$U = 0, Q = q_{+\infty} \quad \text{as } \zeta \to +\infty.$$

Behind this inner wave we match the flux to that of the outer solution

$$\frac{dU}{d\zeta}|_{\zeta\to-\infty} = \frac{d\varphi}{d\xi}|_{\xi\to 0^-},$$

and $Q = 0$ as $\zeta \to -\infty$.
Hence integrating and applying conditions as $\zeta \to \pm\infty$, we obtain

$$u_{-\infty}(\mathbf{v}\cdot\mathbf{n} - N\theta - \varepsilon K) = Nq_{+\infty};$$

which on rearranging gives

$$N\left(\theta + \frac{q_{+\infty}}{u_{-\infty}}\right) = (\mathbf{v}\cdot\mathbf{n} - \varepsilon K). \tag{6.5.4}$$

This is itself a non-linear parabolic equation for the motion of the transition surface M, for our non-planar wave. Note that the term $q_{+\infty}/u_{-\infty}$ is a retardation term due to the reaction providing a sink for the incoming solute species, u, and the motion is curvature damped.

We have obtained an equation (6.5.4) governing the evolution of non-planar wave fronts. For εK close to zero, the front is propagated along each streamline at retarded velocity: hence

$$N = \frac{\mathbf{v}\cdot\mathbf{n}}{\left(\theta + \frac{q_{+\infty}}{u_{-\infty}}\right)},$$

the normal fluid velocity being $\mathbf{v}\cdot\mathbf{n}/\theta$. The term εK may not always be conveniently ignored though, and provides a singular perturbation necessary to smooth out the front, or allow boundary conditions to be imposed: for example, fronts must propagate in such a way as to meet any no-flux boundaries orthogonally as in section 4.3.

Just as in Chapter 4 we have replaced a nonlinear partial differential equation defined over a domain in $\mathbf{R}^3$ with a nonlinear partial differential equation, (6.5.4), for the wave front evolution defined over a domain in $\mathbf{R}^2$, representing the parameterization of the surface M by local coordinates η and λ. In fact, (6.5.4) falls within the class of quasi-linear parabolic systems. This can be seen directly if we choose some explicit coordinates and simplify. For example, we may consider waves propagating within a plane ($\mathbf{R}^2$, say the x-y plane). Here the location of the transition layer,

M, describes a curve. (6.5.4) can be reduced by taking λ to be the orthogonal z direction and imposing symmetry. Hence if the transition layer is parameterized by η, $(x(\eta,t), y(\eta,t))$ say, we have

$$\begin{aligned} N =& \frac{x_t y_\eta - y_t x_\eta}{(x_\eta^2 + y_\eta^2)^{1/2}}, \\ K =& \frac{-x_{\eta\eta} y_\eta + y_{\eta\eta} x_\eta}{(x_\eta^2 + y_\eta^2)^{3/2}}, \\ \mathbf{n} =& \frac{(y_\eta, -x_\eta)^T}{(x_\eta^2 + y_\eta^2)^{1/2}}. \end{aligned}$$

Obviously (6.5.4) is still fairly complicated, but things simplify if we assume that the mapping $\eta \to x$ is invertible, so that the "dummy" coordinate η can be eliminated. Writing $y^*(x,t) = y(\eta,t)$ where $x = x(\eta,t)$, we have

$$\begin{aligned} y_t =& y_x^* x_t + y_t^*, \\ y_\eta =& y_x^* x_\eta. \end{aligned}$$

Now (6.5.4) becomes (on dropping the *s)

$$\left(\theta + \frac{q_{+\infty}}{u_{-\infty}}\right) y_t = \frac{\varepsilon y_{xx}}{(1 + y_x^2)} + \mathbf{v} \cdot \begin{bmatrix} -y_x \\ 1 \end{bmatrix}. \tag{6.5.5}$$

This last equation clearly exposes the form of (6.5.4) as a parabolic system: essentially the curvature correction is a smoothing, dispersive term, limiting the growth of disturbances to the front.

The importance of (6.5.4) in modelling non-planar wave propagation lies in its ability to represent the balance between two processes:

- the heterogeneity of the velocity field, **v**, and
- the role of diffusion in smoothing solutions.

If **v** is variable, then, in the absence of diffusion, the front would propagate along streamlines (as a shock wave) and would develop highly variable structure owing to the differences in travel times.

For example, suppose we have the divergence free field given by

$$\mathbf{v} = \begin{bmatrix} 0 \\ 1 + \mu \cos \omega x \end{bmatrix}$$

for some small μ and real ω, such that $\frac{\mu}{\omega} = O(\varepsilon)$.
We rescale the time axis by introducing

$$\tau = \frac{t}{\left(\theta + \frac{q_{+\infty}}{u_{-\infty}}\right)}, \tag{6.5.6}$$

so that (6.5.5) becomes

$$y_\tau = \frac{\varepsilon y_{xx}}{(1+y_x^2)} + \mathbf{v} \cdot \begin{bmatrix} y_x \\ 1 \end{bmatrix}. \tag{6.5.7}$$

Then, ignoring the curvature,

$$y = \tau(1 + \mu \cos \omega x) + \text{constant} \tag{6.5.8}$$

is a solution, developing figures as a consequence of the variable nature of $\mathbf{v}$. Such a solution is shown in Figure 6.4.

If $\varepsilon \neq 0$, then y_{xx} cannot grow unchecked. In fact for $\varepsilon \neq 0, y$ is asymptotic to a solution of the form

$$y = \tau + \int_0^x \tan\left(-\frac{\mu}{\varepsilon\omega} \sin \omega\tilde{x}\right) d\tilde{x} + \text{constant}. \tag{6.5.9}$$

Comparing (6.5.8) with (6.5.9), we see how the curvature term limits the growth of fingers. Note that (6.5.7) is nonlinear so, in general, cannot be solved explicitly.

In Figure 6.5 we show the x dependence of (6.5.9), which reflects the periodic heterogeneity in $\mathbf{v}$.

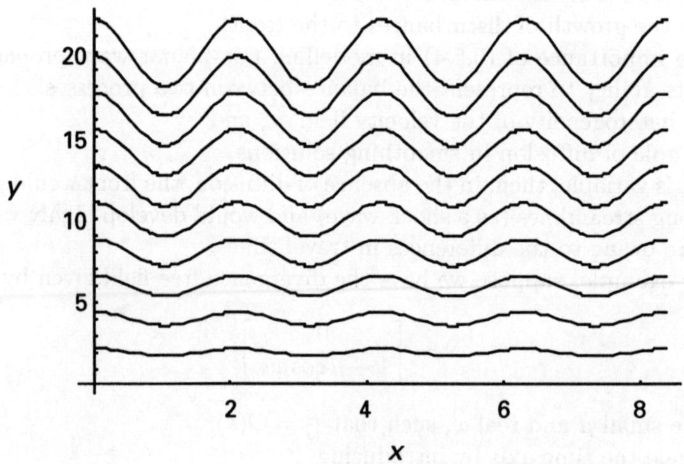

Figure 6.4: In the absence of diffusion, transition front fingers grow continuously, in (6.5.8), with $\mu = 0.1$, $\omega = 3$.

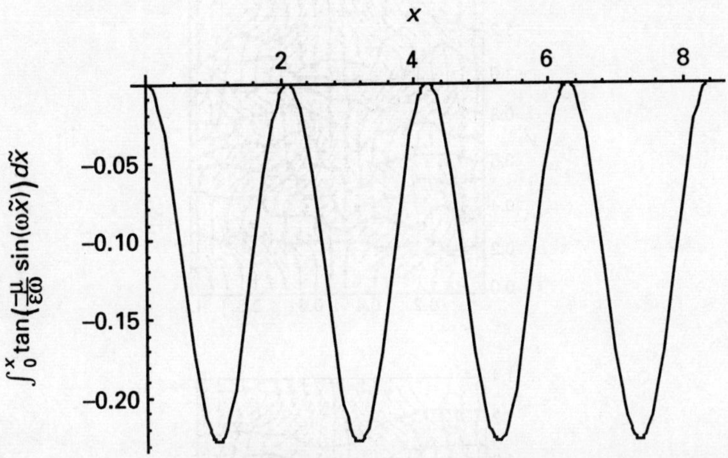

Figure 6.5: In the presence of diffusion-transition front fingers are limited, in (6.5.9), with $\mu = 0.1$, $\omega = 3$ and $\epsilon = 0.1$. Compare with the amplitudes in Figure 6.4.

Expanding out the tan function in (6.5.9), the amplitude of the fingers in (6.5.9) may be written as an asymptotic series in $\mu/\epsilon\omega$. We have oscillations in (6.5.9) of magnitude

$$\frac{1}{\omega}\left\{\frac{\mu}{\omega\varepsilon} + \frac{2\mu^3}{9\omega^3\varepsilon^3} + \frac{16\mu^5}{225\omega^5\varepsilon^5} + O(\mu^7/\varepsilon^7\omega^7)\right\}.$$

Using (6.5.7) we can numerically simulate redox front propagation through highly heterogeneous flow fields.

For example in Figure 6.6 we depict a divergence free field $\mathbf{v} = (x, y)$ generated by specifying an appropriate stream function. Imposing no-flux boundary conditions on the original reaction-diffusion-advection system along the boundaries, at $x = 0$ and $x = 1$, we consider the evolution of a front starting out along the line $y = 0$, which evolves according to (6.5.5), and meets the boundaries orthogonally: hence we solve (6.5.5) together with

$$y_x = 0 \qquad \text{at } x = 0,\ x = 1,\ \text{for } t > 0.$$

Figure 6.6 depicts the solution $y(x, t)$ giving the location of the propagating non-planar front at various fixed times, and for various values of the diffusion parameter, ε.

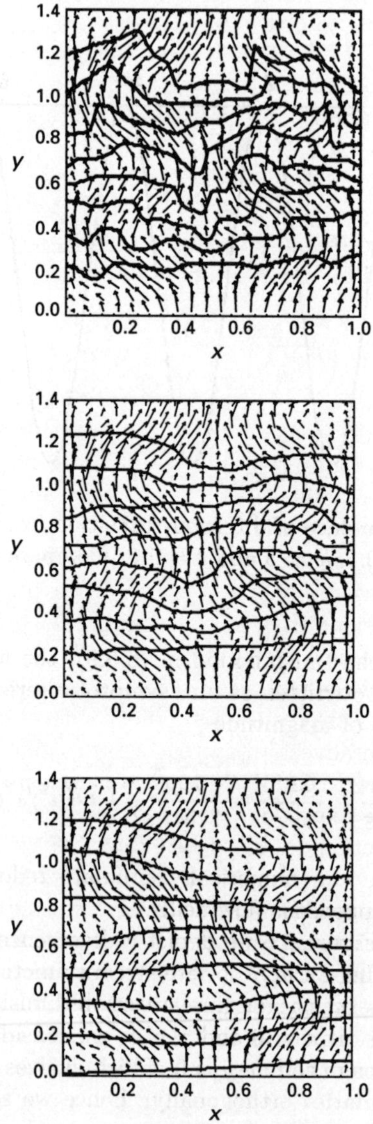

Figure 6.6: Simulation of a curvature-damped transition front propagating through a heterogeneous flow field, satisfying no-flux boundary conditions along $x = 0$ and $x = 1$. The solid lines depict the position of a moving front at successive time increments, as it propagates over the x-y plane, for $\varepsilon = 0.03$ (top), $\varepsilon = 0.3$ (middle) and $\varepsilon = 3$ (bottom).

The units are arbitrary here since time can be rescaled, and the solution only depends upon the balance between the scale of the heterogeneities within the flow field, $\mathbf{v}$, and the size of ε. We choose to vary ε.

In the case of geochemical reactions governed by isotherms (reversible instantaneous distribution between solute and mineral species):

$$\text{solid} = f(u);$$

an equation similar to (6.5.4) can be derived (with only a diffusive transitional layer – no need for the inner reactive layer), in the form

$$N\left(\theta + \frac{f(u_{-\infty})}{u_{-\infty}}\right) = \mathbf{v}\cdot\mathbf{n} - \varepsilon K.$$

Finally we may consider large scale geometrical heterogeneity in assessing the performance of subsurface disposal facilities. In Figure 6.7 we depict the propagation of a reactive front around and through a low permeability engineered barrier, again solving (6.5.4).

We have shown geochemical transport problems lead to the development of successive flux-driven sharp fronts. Their propagation is relatively simple in one dimension: that is, whilst imposing planar symmetry upon the fluid flow and the host medium.

Considerations of the multiple timescales associated with kinetics and dispersive phenomena are required in order to develop the solutions of the full reaction-diffusion-advection system close to the sharp fronts. Nevertheless, we have shown that the existence of the fronts (and their corresponding planar wave speeds) is a robust property effectively determined by conservation of advection dominated solute fluxes either side of the frontal transition profiles.

Although useful in yielding a quantitative understanding of phenomena in one spatial dimension, numerical codes designed for reaction-path problems cannot be extended easily to a higher number of spatial dimensions, especially in the presence of heterogeneity in the underlying flow field. This is because the small dispersive terms play a dual role: not only do they limit the front profiles in the direction of fluid flow, but they also limit the growth of fingering.

In order to analyze such problems, we have developed a geometrical approach to front propagation. Hence the propagation of sharp fronts can be calculated directly as a function of the underlying advective fluid flow, for varying degrees of dispersivity.

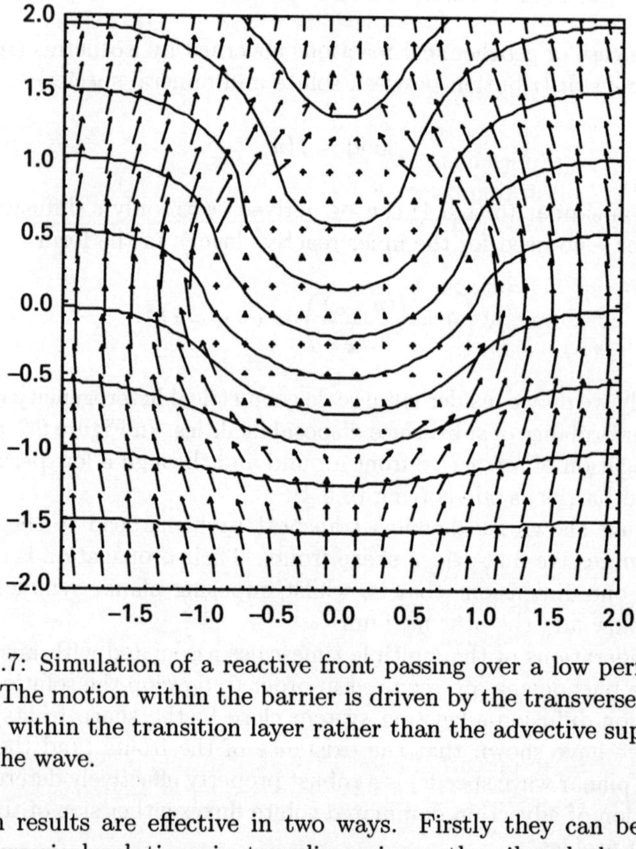

Figure 6.7: Simulation of a reactive front passing over a low permeability barrier. The motion within the barrier is driven by the transverse diffusion of solute within the transition layer rather than the advective supply from behind the wave.

Such results are effective in two ways. Firstly they can be used to verify numerical solutions in two dimensions: the eikonal alternative of "solving for the front" can be compared directly to adaptive or multigrid solutions of the full problem in the presence of natural heterogeneities or spontaneous fingering. Secondly they can be used to estimate, for example, how redox fronts can propagate in and around engineered barriers, which provide geometrically well specified fluctuations in the flow field.

Two generalizations of the results here are worth drawing out. Firstly in considering problems such as those in section 6.4, one obtains successive fronts in one dimension each moving with speed ω_i. As in 6.4 this wave speed can be written in the form

$$\omega_i = \frac{v}{\theta + \mathrm{cap}_i},$$

where cap_i is a measure of the capacity of the medium ahead of the front to retard the wave (e.g.: $= q_{+\infty}/u_{-\infty}$ in (6.5.4)). The corresponding equation of the motion for the non-planar front is thus

$$N = \frac{\omega_i}{|\mathbf{v}|}(\mathbf{v}\cdot\mathbf{n} - \epsilon K).$$

Thus once the one dimensional problem is solved ([4] for example) it is feasible to allow the successive waves to become non-planar and apply the eikonal approach to each successive wave front.

If mineral alteration causes porosity changes, and hence permeability changes, the simulation in two and three dimensions is not so straightforward. In this case the transition surface M divides the domain into two subdomains having differing permeabilities. At each time, in order to calculate the normal component of fluid flux, $\mathbf{v}\cdot\mathbf{n}$, at the surface, one must solve for the flow field $\mathbf{v}$ in both subdomains, matching both fluid pressure and normal fluid flux at M. This is considered in the next section.

6.6 Frontal instability and fingering for fully coupled systems

In a number of papers ([58], [59], [60]) results have been presented concerning fully coupled geochemical systems within saturated porous media, and the ability of near planar waves to develop fingers. Here we follow [13].

Heuristically this phenomena can be understood as follows. At a wavefront a dissolution reaction takes place as the host mineralogy reacts with incoming solute species which are advected towards the reaction zone by the fluid flow. For a fully coupled system this may result in an increase of porosity and of hydrological conductivity. Hence if fingers start to form, as small geometric perturbations to the near planar wave front, the flow field is distorted, seeking the more conductive paths along the fingers, and delivering solute species in greater quantities to the reaction sites at the finger "tips"; hence moving the wave faster at these sites than at the finger "gaps". Of course this process is diffusion limited: if such structures continue to grow then large chemical gradients will be formed. So a balance must be set up, and a stable non-planar wave front propagated even for spatially homogeneous systems.

Mathematically this may be considered as a spontaneous symmetry breaking. There is a plane wave solution which becomes unstable to oscillatory perturbations in the direction traverse to the original motion: we shall consider this as a bifurcation problem.

The purpose of this section is to demonstrate such behaviour for a class of problems where the reaction zone is very small compared with scale of propagation. This is a standard assumption in previous work on fingering [58], and allows us to make use of an important reduction given in the previous section in which we solve for the location of a sharp transition front, rather than the full distribution of chemical species. Effectively, equilibrium holds ahead of, and behind, the front, and its sharpness is dependent upon fast chemical kinetics, and diffusion/dispersion rate which is small compared with the fluid flow rate. Such fronts are propagated along streamlines with curvature corrections, due to diffusion traverse to the direction of propagation, within the front, if sharp structures develop. In the last section we applied this to problems in spatially heterogeneous systems. Here we exploit the same ideas as a starting point for our analysis of fingering for coupled systems.

The results obtained here extend the considerations of [58] since we shall not rely on the assumption that the ratio of incoming solute species (involved in the dissolution reaction) to the density of mineral species (to be dissolved), say δ, is small. Previous results [58], [59], [60] are valid in the limit $\delta \to 0$: yet here this is not required. Moreover we obtain a dispersion curve relating the wave number of the perturbations to the plane wave symmetry to their rate of growth, which incorporates the effect of large δ

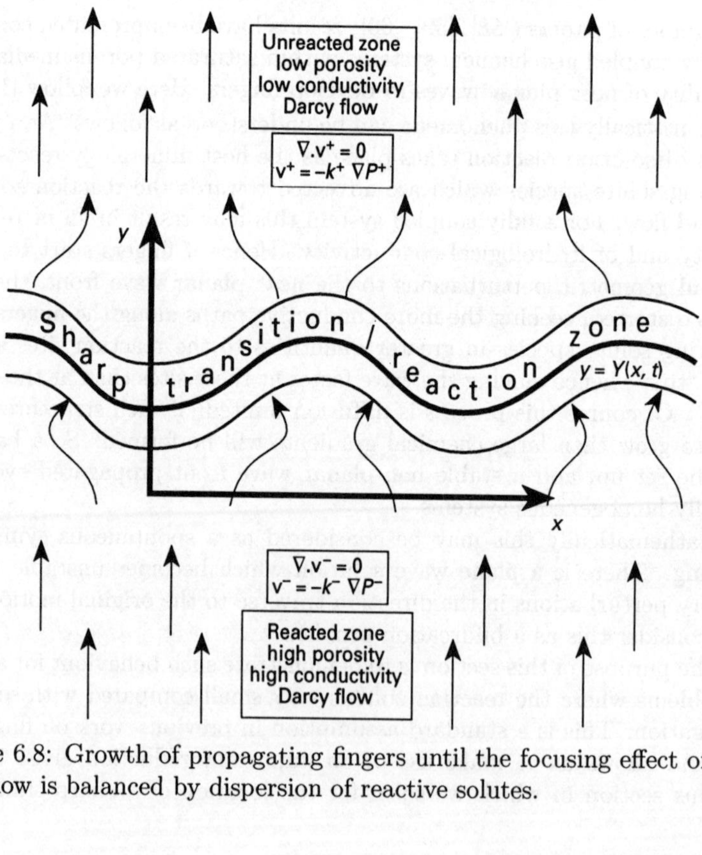

Figure 6.8: Growth of propagating fingers until the focusing effect of coupled flow is balanced by dispersion of reactive solutes.

(where fingering may perhaps be disallowed), and this is consistent with the dominance of the Fickian dispersive term (i.e. is quadratic) for large wave numbers. This extends and generalizes previous results [58], [59]: like them we do not attempt an analysis of the full (unreduced) system of partial differential equations for all of the reactant concentrations in higher dimensions (though plane wave solutions may be accessible, via transition layer analysis). A full result showing fingering instability for such systems remains elusive.

Consider a problem of reactive transport for aqueous chemical species within a saturated porous medium where the incoming fluid is in chemical disequilibrium with the host mineralogy. From section 6.4 such problems result in the development of a sequence of waves, of mineral precipitation and dissolution, each moving at its own characteristic velocity. All such waves are flux driven: that is they propagate as a result of the advective supply of solute species to a thin reaction zone from upstream. Within the zone there are sharp gradients set up in the chemical solute species.

On either side of each such wave the water chemistry is in equilibrium with the local mineralogy. As a consequence of the alteration of the minerals (the precipitation or dissolution of small grains) there is a change in the porosity, and also of the local hydrological conductivity of the medium. This in turn changes the fluid flow field, here assumed to obey Darcy flow for an incompressible fluid, resulting in a change in the advective transport of solute and hence the reaction. As discussed in section 6.1, these systems are referred to as **fully coupled**, as opposed to uncoupled systems where the reaction is assumed not to affect the flow field significantly.

We consider the following problem in the plane, which must be solved for $y = Y(x,t)$, the position of the reactive front, and $P^{\pm}(x,y,t)$, the pressure field ahead of and behind the front, see Figure 6.8. We have

$$(\theta^- + \beta)Y_t = D\frac{Y_{xx}}{1+Y_x^2} + \mathbf{v}^-_{|y=Y} \cdot \begin{bmatrix} -Y_x \\ 1 \end{bmatrix} \tag{6.6.1}$$

$$x \in \mathbf{R},\ t > 0;$$

$$\mathbf{v}^- = -k^-\nabla P^-, \qquad \nabla \cdot \mathbf{v}^- = 0, \tag{6.6.2}$$

$$x \in \mathbf{R},\ y < Y(x,t),\ t > 0;$$

$$\mathbf{v}^+ = -k^+\nabla P^+, \qquad \nabla \cdot \mathbf{v}^+ = 0, \tag{6.6.3}$$

$$x \in \mathbf{R},\ y > Y(x,t),\ t > 0;$$

together with continuity of pressure and conservation of fluid mass at the front:

$$P^+ = P^-, \tag{6.6.4}$$

$$\begin{bmatrix} -Y_x \\ 1 \end{bmatrix} \cdot (\mathbf{v}^- - \mathbf{v}^+) = \frac{Y_t}{(1+Y_x^2)}(\theta^- - \theta^+), \tag{6.6.5}$$

$$x \in \mathbf{R},\ y = Y(x,t),\ t > 0.$$

Here $0 < \theta^+ < \theta^- < 1$ are the (constant) porosities ahead of and behind the reaction front. D denotes the solute dispersion/diffusion rate; $\mathbf{v}^{\pm}(x, y, t)$ denote the Darcy velocity fields ahead of and behind the wave. Equation (6.6.1) incorporates a constant β which represents the retardation of the wave due to the reaction: a plane wave would have normal velocity (in the $\mathbf{n}$ direction)

$$\frac{\mathbf{v}^- \cdot \mathbf{n}}{\theta^- + \beta},$$

whereas the mean normal fluid velocity is $|\mathbf{v}^- \cdot \mathbf{n}| / \theta$. β is solely a function of the incoming chemistry and the host mineralogy. For a simple [solute] + [solid] $\rightarrow$ products reaction (see example at the end of this section) β is given by:

$$\beta = \frac{q_\infty}{u_{-\infty}},$$

where q_∞ is the density of the unreacted mineral and $u_{-\infty}$ is the incoming concentration of the solute (this is analogous to the parameter δ^{-1} in [59], introduced and discussed above).

Equations (6.6.2) and (6.6.3) represent steady Darcy flow for incompressible fluid either side of the front. The conductivities $k^{\pm}$ are constant, such that $0 < k^+ < k^-$. The time dependence of $\mathbf{v}^{\pm}$, and hence $P^{\pm}$, enters via the imposition of the conditions (6.6.4) and (6.6.5) at the moving front $y = Y(x, t)$. Equation (6.6.5) accounts for the difference in the flux of fluid across the front allowing for the change in porosity.

(6.6.1) and (6.6.3) must be solved together with suitable boundary conditions at $x \to \pm\infty$ and uniform flow asymptotically.

$$\mathbf{v}^+ \to \begin{bmatrix} 0 \\ v_{\text{out}} \end{bmatrix} \quad \forall x \in \mathbf{R} \text{ as } y \to \infty$$

$$\mathbf{v}^- \to \begin{bmatrix} 0 \\ v_{\text{in}} \end{bmatrix} \quad \forall x \in \mathbf{R} \text{ as } y \to -\infty \tag{6.6.6}$$

for constants $v_{\text{in}} > 0$, given, and a suitable v_{out}.

A **plane wave** solution exists, given by

$$\mathbf{v}^- \equiv \begin{bmatrix} 0 \\ v_{\text{in}} \end{bmatrix} \; y < Y; \quad \mathbf{v}^+ \equiv \begin{bmatrix} 0 \\ v_{\text{out}} \end{bmatrix} \; y > Y;$$

$$Y(x,t) = \frac{v_{\text{in}}}{\theta^- + \beta} t$$

where $v_{\text{out}} = v_{\text{in}} \left[\frac{\beta+\theta^+}{\beta+\theta^-} \right]$.

The analysis in sections 6.3 and 6.4 can be extended to yield explicit asymptotic representations of this wave profile within the transition layer if required for given reaction terms.

Next following [58] we perturb the solution by introducing a scaling parameter ε and setting

$$Y = \frac{v_{\text{in}} t}{\theta^- + \beta} + \varepsilon e^{\sigma t + i\omega x} + O(\varepsilon^2), \tag{6.6.7}$$

$$P^- = -\frac{v_{\text{in}}}{k^-}\left(y - \frac{v_{\text{in}}}{\alpha} t\right) + \varepsilon A e^{|\omega|(y - \frac{v_{\text{in}}}{\alpha} t)} + O(\varepsilon^2), \tag{6.6.8}$$

$$P^+ = -\frac{v_{\text{out}}}{k^+}\left(y - \frac{v_{\text{in}}}{\alpha} t\right) + \varepsilon B e^{-|\omega|(y - \frac{v_{\text{in}}}{\alpha} t)} + O(\varepsilon^2), \tag{6.6.9}$$

for some constants A and B to be determined, and seeking solutions for ε small which indicate the stability of the wave with respect to such traverse oscillatory perturbations.

Using (6.6.8) and (6.6.9) it is clear that (6.6.2) and (6.6.3) are satisfied together with the limiting conditions (6.6.6). Note that we need only determine the spatial variation in the gradient of the pressure field: addition of an arbitrary function of t to $P^{\pm}$ everywhere makes no contribution. Hence the implicit normalization in (6.6.8) and (6.6.9) so that $P^{\pm} = 0$ at the front as $\varepsilon \to 0$.

Using (6.6.7) and (6.6.8) in (6.6.1) implies

$$(\theta^- + \beta)\sigma = -D\omega^2 - k^- \, |\omega| \, A. \tag{6.6.10}$$

Using (6.6.8) and (6.6.9) at $y = Y(x,t)$ (given in (6.6.7)) the conditions (6.6.4) and (6.6.5) yield

$$B = A + \frac{v_{\text{out}}}{k^+} - \frac{v_{\text{in}}}{k^-}, \tag{6.6.11}$$

$$\sigma(\theta^- - \theta^+) = - \, |\omega| \, (k^- A + k^+ B). \tag{6.6.12}$$

Solving (6.6.10), (6.6.11), (6.6.12) yields σ, A, and B in terms of $|\omega|$, v_{out}, v_{in}, $k^{\pm}$, and $\theta^{\pm}$.

Finally we use $v_{\text{out}} = v_{\text{in}}(\beta + \theta^+)/(\beta + \theta^-)$, to obtain

$$\sigma\left(\beta + \frac{(\theta^- k^+ + \theta^+ k^-)}{(k^- + k^+)}\right) = -D\omega^2 + |\omega|\,\lambda v_{\text{in}}, \tag{6.6.13}$$

where

$$\lambda = \frac{(\beta(k^- - k^+) - k^+\theta^- + k^-\theta^+)}{(\theta^- + \beta)(k^- + k^+)}.$$

For instability, we require $\lambda > 0$, which implies a necessary condition for fingering is

$$\beta > \frac{(k^+\theta^- - k^-\theta^+)}{(k^- - k^+)} \equiv \beta_0, \text{ say.}$$

Note, β_0 can be of either sign.

Fingering will occur when σ is positive, which defines a region $0 < \omega < \omega_0$ where

$$\omega_0 = \frac{\lambda v_{\text{in}}}{D} = \frac{v_{\text{in}}}{D}\left(\frac{\beta(k^- - k^+) - k^+\theta^- + k^-\theta^+}{(\theta^- + \beta)(k^- + k^+)}\right).$$

In Figure 6.9 we depict the curve (6.6.13) for different values of β. The plane wave is stable to traverse perturbations for β small, becoming unstable as β increases through β_0=0.05.

For β asymptotically large, this approaches a result analogous to that of [58]. Indeed, the dispersion curves obtained agree exactly as $\omega \to 0$. However in [58] the behaviour of σ for large ω is asymptotically linear, whereas here it is dominated by the negative quadratic term due to the Fickian dispersion (curvature term) affecting the solute species.

Next we consider a specific illustration of this phenomenon.

Example

Here we follow [13]. As an example of the class of systems discussed in the previous section, here we consider

$$\begin{aligned}
(\theta u)_t &= D\Delta u - \nabla\cdot[\mathbf{v}u] - R \\
q_t &= -R \\
\theta_t + \nabla\cdot\mathbf{v} &= 0 \\
\mathbf{v} &= -k\nabla P \\
(\theta = \theta(q) \quad & k = k(q)) \\
R &= \theta u q/\mu
\end{aligned} \tag{6.6.14}$$

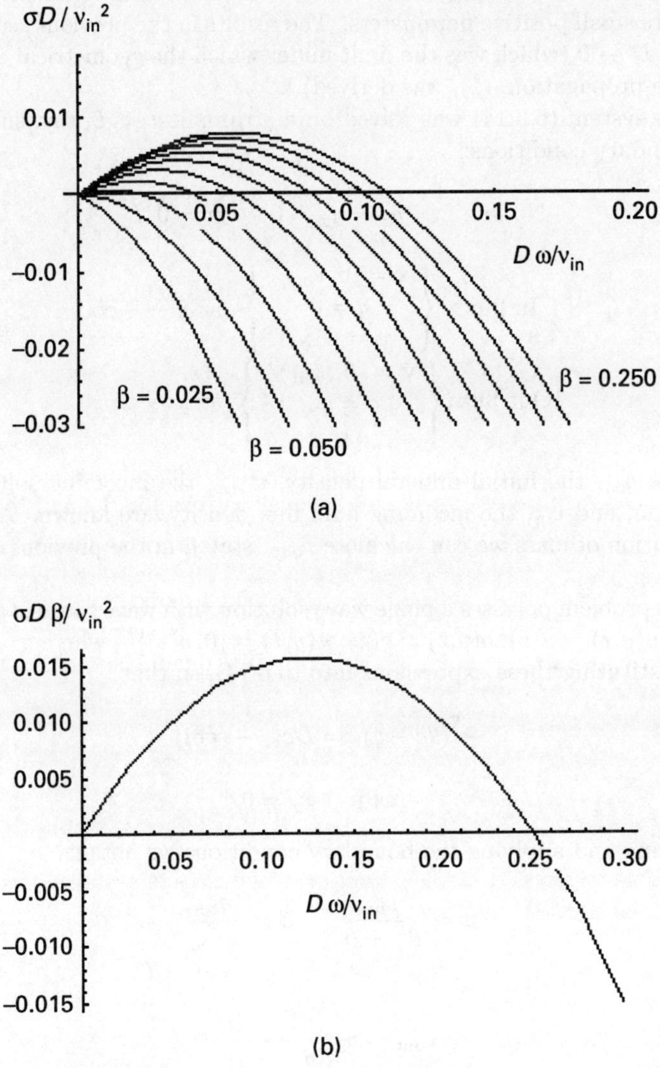

Figure 6.9: Dispersion curves: a) as β increases – in this case the plane wave is stable for $\beta < 0.05$; b) as $\beta \to \infty$. Here we take $k^+ = 6$, $k^- = 10$, $\theta^+ = 0.1$ and $\theta^- = 0.2$.

Here, as usual, $u(x, y, t)$ denotes the concentration of the incoming solute species; $q(x, y, t)$ denotes the concentration of the unreacted mineral within the medium. $\mathbf{v}(x, y, t)$ and $P(x, y, t)$ are the Darcy fluid velocity and the hydrostatic pressure; $\theta(q)$ and $k(q)$ denote the mineralogy dependent porosity and conductivities respectively, both decreasing functions of q. D and μ are small positive parameters. The results in the previous section are valid as $D \to 0$ (which was the limit under which the geometrical equation for wave propagation, (1), was derived).

The system (6.6.14) was solved on a strip $0 < x < L$, say, subject to the boundary conditions:

$$\text{No flux}: \qquad p_x = u_x = 0 \qquad x = 0, L$$

$$\text{In flux} \quad \left\{ \begin{array}{c} \mathbf{v} = (0, v_{\text{in}})^{\text{T}} \\ q = 0 \\ u = u_{-\infty} \end{array} \right\} \quad \text{as } y \to -\infty$$

$$\text{Out flux} \quad \left\{ \begin{array}{c} \mathbf{v} = (0, v_{\text{out}})^{\text{T}} \\ q = q_{\infty} \\ u = 0 \end{array} \right\} \quad \text{as } y \to \infty$$

Here q_∞, the initial mineral density, $u_{-\infty}$, the incoming solute concentration, and v_{in}, the incoming fluid flux density, are known. Applying conservation of mass we can calculate v_{out} (stated in the previous section) as follows.

The problem posesses a plane wave solution with wave speed ω given by writing $u(y, t) = u(z)$, $q(y, t) = q(z)$, $\mathbf{v}(y, t) = (0, v(z))^T$, where $z = y - \omega t$, and substituting these expressions into (6.6.14), so that

$$-\omega(\theta c - q)_z = Dc_{zz} - (vc)_z$$

$$-(\omega\theta)_z + v_z = 0.$$

Integrating and applying the boundary conditions we obtain

$$\omega = \frac{v_{\text{in}}}{\theta^- + \beta}, \qquad \beta = \frac{q_\infty}{u_{-\infty}}$$

and

$$v_{\text{out}} = v_{\text{in}} \frac{(\theta^+ + \beta)}{(\theta^- + \beta)}$$

respectively (here $\theta^- = \theta(0)$ and $\theta^+ = \theta(q_\infty)$). These expressions were stated in the general analysis given above in this section.

The existence of such a travelling solution greatly facilitates the numerical calculation. The position of the reaction front will be given by small perturbations from that calculated analytically for a plane wave solution (relative to the distance the front has travelled). Thus, to proceed numerically we introduce a travelling coordinate $z = y - \omega t$, applying the boundary conditions at fixed distances ahead of and behind such an initial wave front, where the physical conditions are assumed to differ negligibly from those at $\pm\infty$.

We solve the equations for two systems comprising different geometries and initial conditions, the results of which are displayed in Figures 6.10-6.12. The results show contour plots of $(\theta^+ + \theta^-)/2$ where this value is assumed to be indicative of the position of the reaction front.

In the first case we solve the equations for a long, narrow domain, imposing an initial condition of a step change in physical properties along the width. This can be seen most clearly in Figure 6.10. The parameters for the calculation were specially chosen so as to allow for the propagation of a lone finger (the previous analysis implying that other eigenmodes should be stable to small perturbations). Table 6.1 lists the parameters which were used for the calculation. In Figure 6.10 we can see clearly the growth of a single finger and how diffusion limits this growth to a steadily propagating finger (convergence of the solution in the moving frame).

Figure 6.11 illustrates the same situation when viewed in a fixed coordinate system (for early times).

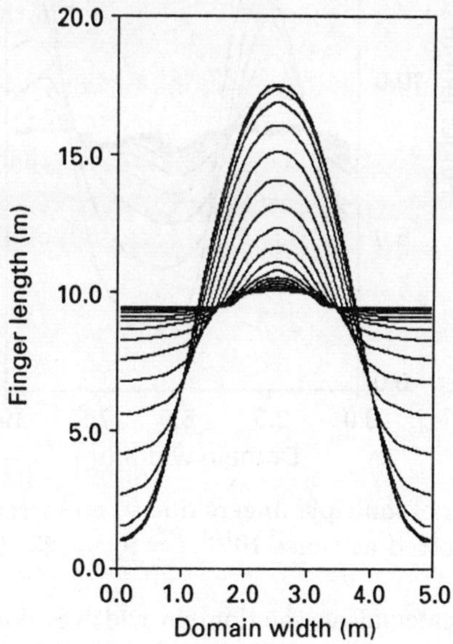

Figure 6.10: Growth of a lone finger from a small initial perturbation as viewed in a travelling coordinate system. The contours are plotted at times $10^{i/4}$, $i = 9, \ldots, 31$ [13].

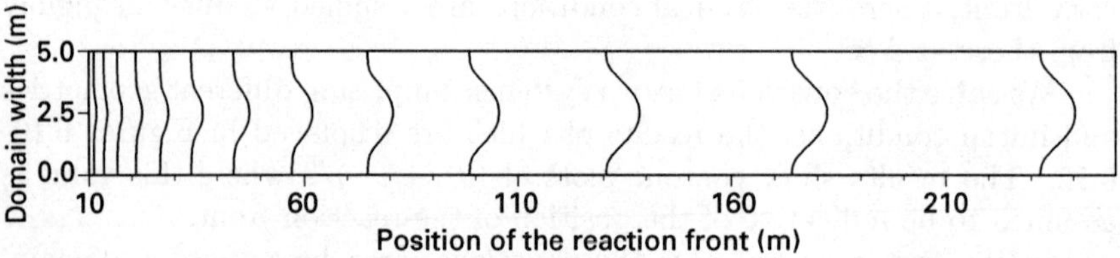

Figure 6.11: Growth of the finger in Figure 6.10 as viewed in a fixed coordinate system. The contours are plotted at times $10^{i/4}, i = 12, \ldots, 27$ [13].

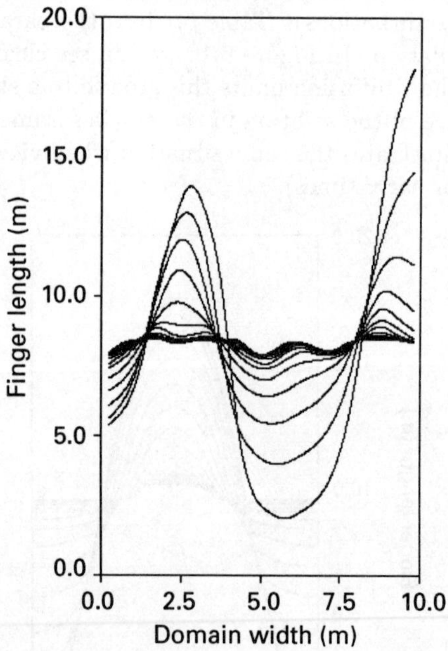

Figure 6.12: Growth of multiple fingers due to an increased domain width. The contours are plotted at times $10^{i/4}, i = 9, \ldots, 23$ [13].

For the second calculation the domain width is doubled (10.0 m) and small random perturbations are applied to the initial plane wave front. The results can be seen in Figure 6.12. As expected, the larger domain width allows for a greater number of unstable eigenmodes, increasing the solution complexity. The behaviour of these solutions compares well with that predicted by the analysis.

Parameter	Value
$u_{-\infty}$	1.0
$q_{+\infty}$	10.0
v_{in}	1.0
D	0.1
μ	0.01
θ^-	0.2
θ^+	0.1
k^-	10
k^+	6

Table 6.1: Parameters used for the calculations shown in Figures 6.10, 6.11 and 6.12.

Notation

Here is a brief summary of the notation and conventions used throughout the text.

x Position vector in $\mathbf{R}^n$ specified in Cartesian coordinates.

r Position vector in $\mathbf{R}^n$ specified in general coordinates: Cartesian, spherical or general curvilinear.

Ω A simply connected open subset of $\mathbf{R}^n$, used to indicate the domains of functions and equations.

∇ The operator nabla, given by $\mathbf{i}\frac{\partial}{\partial x} + \mathbf{j}\frac{\partial}{\partial y} + \mathbf{k}\frac{\partial}{\partial z}$ in the usual Cartesian (x, y, z) coordinates, having orthonormal basis $\mathbf{i}, \mathbf{j}, \mathbf{k}$. Operating on scalar fields, ∇f is the gradient of f. Operating on vector fields, $\nabla.\mathbf{u}$ and $\nabla \times \mathbf{u}$ are the divergence and curl of $\mathbf{u}$, respectively.

Δ The Laplacian operator $\nabla.\nabla = \frac{\partial^2}{\partial x^2} + \frac{\partial^2}{\partial y^2} + \frac{\partial^2}{\partial z^2}$. This is written as ∇^2 in some texts.

$C^0(\Omega, X)$ The space of all uniformly bounded continuous functions on Ω, taking values in the space $X = \mathbf{R}^n$ or $\mathbf{C}^n$. This function space is a Banach space, that is a complete, normed vector space, equipped with the supremum (or infinity) norm

$$\|f\| = \sup_{x \in \Omega} \left\{ \|f(x)\|_X \right\}.$$

$C^k(\Omega, X)$ The space of those functions in $C^0(\Omega, X)$ which possess derivatives of order $\leq k$ which also lie in $C^0(\Omega, X)$. It has the natural norm

$$\|f\| = \sum_{j=0}^{k} \sup_{x \in \Omega} \left\{ \|\partial_j f\|_X \right\},$$

where we allow $\|\partial_j f\|_X$ to denote an implicit sum over the norms of all jth order derivatives of f.

$L_p(\Omega, X)$ The space of all p^{th} power integrable functions on Ω, taking values in the space X, which will always be either $\mathbf{R}^n$ or $\mathbf{C}^n$. This function space is a Banach space equipped with the norm

$$\|f\| = \left(\int_\Omega \|f(x)\|_X^p \, dx \right)^{1/p}.$$

Where the range is obvious from the context we shall write $L_p(\Omega)$, rather than say $L_p(\Omega, \mathbf{R})$.

$L_2(\Omega, \mathbf{C})$ A special case, which is a Hilbert space, that is, its norm is subordinate to the inner product

$$< f, g >= \left(\int_\Omega f g^* \, dx \right),$$

where * denotes complex conjugation. More generally $L_2(\Omega, \mathbf{C}^n)$ is a Hilbert space, relying upon the inner product on $\mathbf{C}^n$.

$H_k(\Omega)$ The Sobolev space of those functions in $L_2(\Omega)$ possessing derivatives of order $\leq k$ also in $L_2(\Omega)$. It has the natural norm

$$\|f\| = \left(\int_\Omega \sum_{j=0}^{k} \|\partial_j f(x)\|_X^p \, dx \right)^{1/p},$$

where, again, $\|\partial_j f\|_X$ denotes an implicit sum over the norms of all jth order derivatives of f. $H_k(\Omega)$ is also a Hilbert space using the inner product induced by that on $L_2(\Omega)$. In fact the Sobolev spaces can be generalized further by using a Fourier transform type of approach to define $H_s(\Omega)$ for all $s \in \mathbf{R}$. In general, the larger the value of s (or k), the smoother the functions – they possess an increasing number of derivatives, in the sense of distributions.

Bibliography

[1] D.G. Aronson, Bifurcation phenomena associated with nonlinear diffusion, in: *Partial Differential Equations and Dynamical Systems* (ed; W.E. Fitzgibbon III), Research Notes in Math. 101, Pitman, London, 1984.

[2] D.G Aronson, L.A. Peletier, Large time behaviour of solutions of the porous media equation in bounded domains, *J. Differ. Equations* **39**, 378-412, 1981.

[3] D.G. Aronson, H.F. Weinberger, Nonlinear diffusion in population genetics, combustion and nerve pulse propagation, in: *Partial Differential Equations and Related Topics* (ed; J.A. Goldstein), Lecture Notes in Math. 446, Springer-Verlag, New York, 1975.

[4] C. Bäverman, Development of "CHEMFRONTS", a coupled transport and geochemical program to handle reaction fronts, in: *SKB Technical Report 93-2*, 1993.

[5] G. Birkoff, G.-C. Rota, *Ordinary Differential Equations* , Wiley, New York, 1978.

[6] G.A. Carpenter, A geometric approach to singular perturbation problems with applications to nerve impulse equations, *J. Differ. Equations* **23**, 1977.

[7] J. Carr, *Applications of Centre Manifold Theory*, Applied Mathematical Sciences, 35, Springer-Verlag, New York, 1981.

[8] S.N. Chow, J.K Hale, *Methods of Bifurcation Theory*, Springer-Verlag, New York, 1982.

[9] K.N. Chueh, C.C. Conley, J.A. Smoller, Positive invariant regions for systems of nonlinear diffusion equations, *Indiana Univ. Math.* **26**, 335-367, 1977.

[10] C.C. Conley, *Isolated Invariant Sets and the Morse Index*, Conference Board of the Mathematical Sciences 38, AMS, Providence R.I., 1978.

[11] E.D. Conway, Diffusion and the predator-prey interaction: patterns in closed systems, in: *Partial Differential Equations and Dynamical Systems* (ed; W.E. Fitzgibbon III), Research Notes in Math. 101, Pitman, London, 1984.

[12] H.S.M. Coxeter, *Introduction to Geometry*, Wiley, New York, 1961.

[13] S.P. Crompton and P. Grindrod, A geometric approach to fingering instabilities for reaction fronts in fully coupled systems, to appear, *IMA Journal of Applied Mathematics*, 1996.

[14] J.N. Dewynne, A.C. Fowler, P.S. Hagan, Multiple reaction fronts in the oxidation-reduction of iron-rich uranium ores, *SIAM Journal of Applied Mathematics* **53**, 971-989, 1993.

[15] H. Engler, Relations between travelling wave solutions of quasilinear parabolic equations, *Proc. Am. Math. Soc.* **93**, 297-302, 1985.

[16] P.C. Fife, Boundary and internal transition layer phenomena for pairs of second order equations, *J. Math. Anal. Appl.* **54**, 497-521, 1976.

[17] P.C. Fife, Stationary patterns for reaction-diffusion equations, in: *Nonlinear Diffusion*, Research Notes in Math. **14**, Pitman, London, 1977.

[18] P.C. Fife, *Mathematical Aspects of Reacting and Diffusing Systems*, Lecture Notes in Biomath. 28, Springer-Verlag, New York, 1979.

[19] P.C. Fife, G.S. Gill, The phase-field desciption of mushy zones, in: *Proc. Conf. on Nonlinear Partial Differential Equations,* Provo, 1987.

[20] P.C. Fife, M.M. Tang, Comparison principles for reaction-diffusion systems, *J. Differ. Equations* **40**, 168-185, 1981.

[21] R. FitzHugh, Mathematical models of excitation and propagation in nerve fibres, in: *Biological Engineering* (ed; H.P. Schwan), McGraw-Hill, New York, 1969.

[22] G.B. Folland, *Introduction to Partial Differential Equations*,Princeton Univ. Press, 1976.

[23] M. Golubitsky, D.G. Schaeffer, *Singularities and Groups in Bifurcation Theory* Vol. 1, Spinger-Verlag, New York, 1985.

[24] J. Gomatam, P. Grindrod, Three dimensional waves in excitable reaction-diffusion systems, *J. Math. Biol.* **25**, 611-622, 1987.

[25] P. Grindrod, J. Gomatam, The geometry and motion of reaction-diffusion waves on closed two dimensional manifolds, *J. Math. Biol.* **25**, 597-610, 1987.

[26] P. Grindrod, Y Hosono, Travelling front solutions for clustering models in population dynamics, preprint, to appear.

[27] P. Grindrod, Models of individual aggregation in single and multi-species communities, *J. Math. Biol.* **26**,651-660, 1988.

[28] P. Grindrod, S. Sinha, J.D. Murray, Steady-state patterns in a cell-chemotaxis model, *IMA J. Math. Appl. in Med. Biol.*, **6**, 69-79, 1989.

[29] P. Grindrod, B.D. Sleeman, Qualitative analysis of reaction diffusion systems modelling coupled unmyelinated nerve axons, *IMA J. Math. Appl. Med. Biol.* **1**, 289-307, 1984.

[30] P. Grindrod, B.D. Sleeman, Comparison principles in the analysis of reaction-diffusion systems modelling unmyelinated nerve fibres, *IMA J. Math. Appl. Med. Biol.* **1**, 343-363, 1984.

[31] P. Grindrod, B.D. Sleeman, Weak travelling fronts for population models with density dependent dispersion, *Math. Meth. Appl. Sci.* **9**, 576-586, 1987.

[32] P. Grindrod, Some reactive transport, dispersal and flow problems associated with geological disposal of radioactive waste, in: *Ordinary and Partial Differential Equations* IV, (ed. B.D. Sleeman and R.J. Jarvis), Longman, 1993.

[33] M.E. Gurtin, R.C. MacCamy, On the diffusion of biological populations, *Math. Biosci.* **33**, 35-49, 1977.

[34] K.P. Hadeler, Travelling fronts and free boundary value problems, *Conf. on Free Boundary Value Problems, Oberwolfach 1981, Proc.* (ed; J. Albrecht, L. Collatz, K.H. Hoffmann), Birkhauser-Verlag, 1982.

[35] D. Henry, *Geometrical Theory of Semilinear Parabolic Equations*, Lecture Notes in Math. 840, Springer-Verlag, New York, 1981.

[36] A.L. Hodgkin, A.F. Huxley, A qualitative description of membrane current and its application to conduction and excitation in nerves, *J. Physiol.* **117**, 500-544, 1952.

[37] W.F. Hughes, J.A. Brighton, *Theory and Problems of Fluid Dynamics*, Schaum's Outline Series, MacGraw-Hill, New York, 1967.

[38] T. Kato, *Perturbation Theory for Linear Operators*, Springer-Verlag, New York, 1976.

[39] J.P.Keener, A geometrical theory for spiral waves in excitable media, *SIAM J. Appl. Math.* **46**, 1039-1056, 1986.

[40] J.P. Keener, J.J. Tyson, Spiral waves in the Belousov-Zhabotinsky reaction, *Physica* **21D**, 307-324, 1986.

[41] D.L. Kreider, R.G. Kuller, D.R. Ostberg, F.W. Perkins, *An Introduction to Linear Analysis*, Addison-Wesley, 1966.

[42] E. Kreyzig, *Intoduction to Functional Analysis With Applications*, Wiley, New York, 1978.

[43] D.A. Lauffenberger, Chemotaxis and cell aggregation, in: *Modelling Patterns in Space and Time* (eds; W. Jager, J.D. Murray), Lecture Notes in Biomath. 55, Springer-Verlag, New York, 1984.

[44] R.J. LeVeque, Numerical Methods for Conservation Laws, Lectures in Mathematics ETH Zurich, Birkhäuser, Berlin, 1990.

[45] S.A. Levin, *Population models and community structure in heterogeneous environments, in: Studies in Mathematical Biology*, Part II (ed; S.A. Levin), M.A.A. Studies in Mathematics, 1978.

[46] A.C. MacBride, *Semigroups of Linear Operators: An Introduction*, Research Notes in Math 156, Pitman, London 1987.

[47] G. de Marsily, *Qualitative Hydrology*, Academic Press, 1986.

[48] H.P. McKean, Nagumo's equation, *Adv. in Math.* **4**, 209-233, 1970.

[49] H. Meinhardt, Models for the ontogenetic development of higher organisms, *Rev. Physiol. Biochem. Pharmacol.* **80**, 47-104, 1978.

[50] M. Mimura, M. Tabata, Y. Hosono, Multiple solutions of two point boundary value problems of Neumann type with a small parameter, *SIAM J. Math. Anal.* **11**, 613-631, 1980.

[51] S. Mizohata, *The Theory of Partial Differential Equations*, Camb. Univ. Press, Cambridge, 1973.

[52] J.D. Murray, *Asymptotic Analysis*, Springer-Verlag, Berlin, 1984.

[53] J.D. Murray, G.F. Oster, Generation of biological pattern and form, *IMA J. Math. Appl. Med. Biol.* **1**, 51-75, 1984.

[54] J.D. Murray, G.F. Oster, Cell traction models for generating pattern and form in morphogenesis, *J. Math. Biol.* **19**, 265-279, 1984.

[55] J. Nagumo, S. Arimoto, S. Yoshizawa, An active pulse transmission line simulating nerve axon, *Proc. IRE* **50**, 2061-2070, 1962.

[56] I. Neretnieks, C. Nyman, Use of coupled geochemical and transport calculations for nuclear waste problem, in: *High Level Radioactive Waste Management, Proceedings of the Fourth Annual International Conference* **2**, pp. 1533-1538, 1993.

[57] A. Okubo, *Diffusion and Ecological Problems: Mathematical Models*, Biomathematics 10, Springer-Verlag, New York, 1980.

[58] P. Ortoleva, E. Merino, J. Chadam and A. Sen, Geochemical self-organization II: the reactive-infiltration instability, *Am J. Sci.*, **287**, 1008-1040, 1987.

[59] W. Chen and P. Ortoleva, Reaction front fingering in carbonate cemented sandstones, in: *Self-Organization in Geological Systems: Proceedings of a Workshop, 26-30 June 1988*, University of California Santa Barbara (ed. P. Ortoleva, B. Hallet, A. McBirney, I. Meshri, R. Reeder and P. Williams). *Earth Science Reviews* **29**, 183-198, 1990.

[60] P. Ortoleva, W. Chen, *Development of Complex Reaction Front Morphologies Through Non-Linear Fluid-Rock Interaction: Modelling, Asymptotics and Numerical Studies*, Modelling and Analysis of Diffusive and Advective Processes in Geosciences, SIAM, 1992.

[61] G.F. Oster, J.D. Murray, Pattern formation models and developmental constraints, to appear.

[62] A. Papoulis, *Probability, Random Variables, and Stochastic Processes*, McGraw-Hill, New York, 1965.

[63] A. Pazy, Semigroups of Linear Operators and Applications to Partial Differential Equations, Applied Mathematical Sciences 44, Springer-Verlag, New York, 1983.

[64] L.A Peletier, The porous medium equation, in:*Applications of Nonlinear Analysis* (ed; H. Amann, N. Bazley, K. Kirchgrassner), Pitman, London, 1981.

[65] J. Rauch, J.A. Smoller, Qualitative theory of the FitzHugh-Nagumo equations, *Adv. Math.* **27**, 12-44, 1978.

[66] H. Risken, *The Fokker-Planck Equation*, Springer-Verlag, Berlin, 1984.

[67] J. Rinzel, J.B. Keller, Travelling wave solutions of a nerve conduction equation, *Biophys. J.* **13**, 1973.

[68] J. Roughgarden, *Theory of Population Genetics and Environmental Ecology: An Introduction*, Macmillan, New York, 1979.

[69] L.A. Segel, *Mathematical Models in Molecular and Cellular Biology*, Camb. Univ. Press, Cambridge, 1980.

[70] J.A. Smoller, *Shock Waves and Reaction Diffusion Equations*, Springer-Verlag, New York, 1983.

[71] J.A. Smoller, A. Tromba, A. Wasserman, Nondegenerate solutions of boundary value problems, *Nonlinear Anal.* **4**, 207-216, 1980.

[72] J.A. Smoller and A. Wasserman, Global bifurcation of steady-state solutions, *J. Differ. Equations* **39**, 269-290, 1981.

[73] D. Terman, Comparison theorems for reaction-difusion systems defined in an unbounded domain, Univ. of Wisconsin, Madison Mathematics Research Center, Technical Summary Report 2374, 1982.

[74] A.B. Tayler, *Mathematical Models in Applied Mechanics*, Oxford Univ. Press, Oxford, 1986.

[75] A.M. Turing, The chemical basis of morphogenesis, *Philos. Trans. R. Soc.* B **237** (37), 37-72, 1952.

[76] B.J. Welsh, J. Gomatam, A. Burgess, Three dimensional chemical waves in the Belousov-Zhabotinsky reaction, *Nature* **304**, 611-614, 1983.

[77] A.T Winfree, *When Time Breaks Down*, Princeton Univ. Press, 1987.

[78] A.T. Winfree, S.H. Strogatz, Singular filaments organise chemical waves in three dimensions: 4. wave taxonomy, *Physica* **13D**, 221-233, 1984.

[79] V.S. Zykov, Modelling of Wave Processes in Excitable Media (in Russian), Nauka, Moscow, 1984 (Eng. Trans. in prep., ed. A.T. Winfree, Manchester Univ. Press).

Index